Oklahoma Notes

Basic-Sciences Review for Medical Licensure
Developed at
The University of Oklahoma, College of Medicine

Suitable Reviews for:
United States Medical Licensing Examination
(USMLE), Step 1
Federation Licensing Examination (FLEX)

Oklahoma Notes

Neuroanatomy

Third Edition

P.A. Roberts

Springer-Verlag
New York Berlin Heidelberg London Paris
Tokyo Hong Kong Barcelona Budapest

P.A. Roberts, Ph.D.
Department of Anatomical Sciences
Health Sciences Center
The University of Oklahoma at Oklahoma City
Oklahoma City, OK 73190
USA

Library of Congress Cataloging-in-Publication Data
Roberts, P.A. (Philip Alex), 1934
    Neuroanatomy / P.A. Roberts.—3rd ed.
        p.   cm.—(Oklahoma notes)

        ISBN-13: 978-0-387-97777-5   e-ISBN-13: 978-1-4612-2902-5
        DOI: 10.1007/978-1-4612-2902-5

        1. Neuroanatomy.   I. Title.   II. Series.
        [DNLM: 1. Nervous System—anatomy & histology.   WL 101
    R646n]
    QM451.R59   1992
    611'.8—dc20
    DNLM/DLC
    for Library of Congress   91-5248

Printed on acid-free paper.

Production managed by Henry Krell; manufacturing supervised by Jacqui Ashri.
Camera-ready copy prepared by the author.

9 8 7 6 5 4 3 2 1

Preface to the
Oklahoma Notes

In 1973, the University of Oklahoma College of Medicine instituted a requirement for passage of the Part 1 National Boards for promotion to the third year. To assist students in preparation for this examination, a two-week review of the basic sciences was added to the curriculum in 1975. Ten review texts were written by the faculty: four in anatomical sciences and one each in the other six basic sciences. Self-instructional quizzes were also developed by each discipline and administered during the review period.

The first year the course was instituted the Total Score performance on National Boards Part I increased 60 points, with the relative standing of the school changing from 56th to 9th in the nation. The performance of the class since then has remained near the national candidate mean (500) with a range of 467 to 537. This improvement in our own students' performance has been documented (Hyde et al: Performance on NBME Part I examination in relation to policies regarding use of test. J. Med. Educ. 60:439–443, 1985).

A questionnaire was administered to one of the classes after they had completed the Boards; 82% rated the review books as the most beneficial part of the course. These texts were subsequently rewritten and made available for use by all students of medicine who were preparing for comprehensive examinations in the Basic Medical Sciences. Since their introduction in 1987, over a quarter of a million copies have been sold. Assuming that 60,000 students have been first-time takers in the intervening five years, this equates to an average of four books per examinee.

Obviously these texts have proven to be of value. The main reason is that they present a *concise overview* of each discipline, emphasizing the content and concepts most appropriate to the task at hand, i.e., passage of a comprehensive examination over the Basic Medical Sciences.

The recent changes in the licensure examination structure that have been made to create a Step 1/Step 2 process have necessitated a complete revision of the Oklahoma Notes. This task was begun in the summer of 1991; the book you are now holding is a product of that revision. Besides bringing each book up to date, the authors have made every effort to make the texts and review questions conform to the new format of the National Board of Medical Examiners tests.

I hope you will find these review books valuable in your preparation for the licensure exams. Good Luck!

Richard M. Hyde, Ph.D.
Executive Editor

Preface to the Third Edition

Three additional sections have been added in this edition in response to students' suggestions. A short summary of motor deficits, a brief discussion of cranial nerve functions and some examples of localizing signs of anatomic lesions should prove helpful in applying basic principles to clinical situations.

P.A. Roberts

Preface

The following notes, originally known as *Neuroanatomical Notations*, were prepared for the purpose of providing a quick review of some of the pertinent points that should be considered in refreshing your memory of Neuroanatomy.
Obviously the booklet is not encyclopedic, and is certainly not intended as a text on the subject. However, hopefully, it will serve as a useful guide and be of aid in the task of systematically preparing for Part I of the National Boards and similar examinations.

P.A. Roberts

"I can't believe that!" said Alice.

"Can't you!" The Queen said in a pitying tone.

"Try again: Draw a long breath, and shut your eyes."

Alice laughed, "There's no use trying," she said.

"One can't believe impossible things."

"I dare say you haven't had much practice," said the Queen.

"When I was your age, I always did it for half an hour a day. Why, sometimes I've believed as many as six impossible things before breakfast . . ."

From: *Through the Looking Glass*
by Lewis Carroll

Contents

CELLULAR COMPOSITION OF NERVE TISSUE

Nerve tissue consists of numerous varieties of neurons which are structurally and functionally supported by glial cells of several different types. Other associated cells include ependymal cells (lining ventricular cavities), satellite cells (sensory and autonomic ganglia), and Schwann cells (peripheral nerve myelin).

NEURONS

Neurons are comprised of the nucleated cell body with cytoplasmic processes extending from it, and may be classified according to the number of processes possessed.

Thus: Unipolar Neurons - 1 process (e.g. sensory ganglia in the peripheral nervous system).
      Bipolar Neurons - 2 processes (special senses)
      Multipolar Neurons - more than 2 processes (central nervous system and postganglionic cells of autonomic nervous system)

NOTE: Unipolar neurons develop from bipolar cells in the embryo. This occurs by fusion of the processes as seen below.

Unipolar Neuron Development

Functionally, neuron processes may be axons, which conduct nerve impulses away from the cell body, or dendrites, transmitting nerve impulses towards the cell body.

Multipolar neurons vary enormously in size and shape, and in the number and arrangement of their dendrites, etc.

Hence: Multipolar neurons may be described as spherical, stellate, conical, pyramidal, flask-shaped, irregular, etc. Size may vary from 5 microns to 100 microns.

      Dendrites may possess small spines (gemmules) which greatly increase the surface area available for synaptic contact. Purkinje cells in the cerebellum may possess up to 60,000 spines.

Golgi Type I Neurons - possess long axons
Golgi Type II Neurons - possess short axons

HISTOLOGICAL STRUCTURE OF NEURONS

The cell body possesses cytoplasm with inclusions and various organelles, and usually one nucleus. Dendrites and axons are formed by extension of the cytoplasm away from the cell body.

Nucleus

The neuron nucleus usually has a distinct nuclear membrane and generally contains a prominent nucleolus (sometimes two nucleoli) consisting of dense aggregations of R.N.A. The nucleolus stains prominently with Toluidin Blue or Thionine.

Cytoplasm

May be abundant as seen in large multipolar motor neurons in the spinal cord, or sparse (e.g., Horizontal cells of the cerebral cortex).

Nissl Substance

Clumps of ribonucleoprotein scattered throughout the cytoplasm, which are demonstrated with Toluidin Blue or Thionine, are unique to nerve cells. Nissl substance is found in the cytoplasm of dendrites and the cell body, but is not present in the axon or axon hillock (point where axon attaches to cell body). Nissl substance is usually more coarse and prominent in motor neurons, whereas in sensory neurons it is finely dispersed and dust-like.

Dissolution of Nissl substance occurs following injury to a nerve cell (Chromatolysis). This may also occur after a prolonged period of excessive stimulation of a nerve cell.

Neurofibrils

With appropriate silver stains, fine filaments may be identified in the cell body cytoplasm extending into the entire length of the axon and its branches. Thought to be functional in forming neurotubules for conducting material from the cell body to the termination of the axon (and vice versa). Thought to be artefacts by some investigators but tissue culture seems to support the real nature of these structures.

Mitochondria

These organelles are found in dendrites, cell body cytoplasm and axon cytoplasm. They are particularly numerous at the terminal end of axons and at the nodes of Ranvier.

Golgi Apparatus

Usually located fairly close to the nucleus, and comprised of a membrane network somewhat more coarse in structure than the neurofibrils. Product storage function may be related to neurosecretory activity of neurons.

Other Cytoplasmic Inclusions

The following entities may also be observed in neuron cytoplasm - vacuoles, glycogen, iron, melanin pigment (e.g. Substantia Nigra cells), lipochrome pigment (particularly in older nerve cells), neurosecretory granules (e.g. in cells of the supraoptic and paraventricular nuclei of the hypothalamus, containing Oxytocin and Anti Diuretic Hormone).

Axon

The axis cylinder containing axoplasm, neurofibrils and some mitochondria, and is contained by the axolemma, a limiting membrane at surface about 300 A in thickness.

Myelin Sheath

Surrounds the axolemma and is discontinuous at intervals, where the nodes of Ranvier are found.

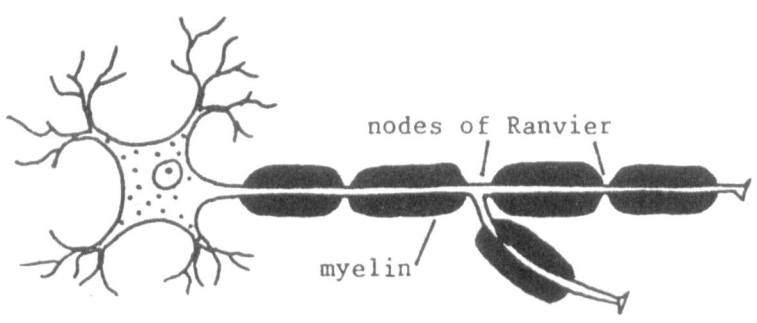

Multipolar Neuron showing Myelin Sheath

The region of myelin found between the nodes of Ranvier, the internode, is usually the product of one Schwann cell, and its nucleus may be seen at the periphery of the myelin sheath. Thinner nerve fibers have shorter internodes than larger diameter fibers. In the central nervous system the nodal gaps are larger than they are in fibers of the peripheral portions of the nervous system.

Myelin is comprised of cholesterol, phospholipids and cerebrosides, wrapped around the axis cylinder in laminated fashion (jelly roll).

In the peripheral nervous system myelin is produced by Schwann cells.

NOTE: Central nervous system myelin is formed by oligodendrocytes
           rather than Schwann cells.

Schmidt-Lanterman clefts are apparent defects in the myelin sheath due to regions possessing excessive amounts of Schwann cell cytoplasm.

Non-Myelinated Nerve Fibers

Some nerve fibers are not invested with an obvious myelin sheath as described above. However, these so called non-myelinated fibers have a close relationship with Schwann cells.

These nerve fibers can be seen to be surrounded by the cytoplasm and cell membrane of Schwann cells, which occasionally invaginate several nerve fibers.

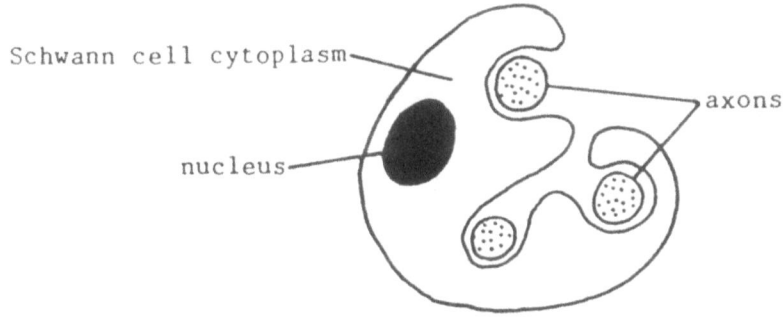

Relationship of Schwann Cell and Axons

Axon Terminals

A short distance prior to termination of the axon, the myelin sheath is lost. The axon terminates as a rounded expansion (Bouton), upon the cell membrane of the structure being innervated (cell body of another neuron, dendrite, axon, or muscle fibers).

Neurotransmitter substance (frequently acetyl choline) is released into the cleft upon arrival of nerve impulse, and acts upon receptor sites found on the subsynaptic membrane.

Other neurotransmitter substances include dopamine, norepinephrin, gamma amino butyric acid, glycine and seratonin.

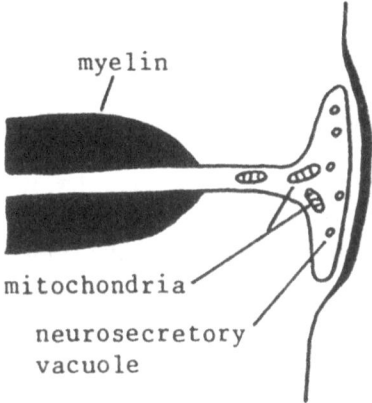

Subsynaptic membrane of the innervated cell, separated from the axon by a subsynaptic cleft 200 Å wide.

Diagram of Synaptic Terminal

A similar arrangement is seen in motor end plates (neuro muscular terminals in skeletal muscle).

GLIAL CELLS

<u>Astrocytes</u> (Largest Glial Cells)

As the name implies these glial cells are star-shaped and possess numerous processes.
They form a sub-ependymal glial membrane beneath the ependymal cells on ventricular surfaces, and also part of the pial-glial membrane on the outer surface of the C.N.S.

Astrocytes are of two basic types, <u>protoplasmic</u> astrocytes containing abundant cytoplasm and found in large numbers in the grey matter, and <u>fibrous</u> astrocytes with smaller amounts of cytoplasm which are more numerous in the white matter of the C.N.S.

Both astrocyte types may possess processes which end in close proximity to blood vessels in the C.N.S. as well as processes in close relation to nerve cells.

> Astrocytic reaction to injury - Swelling - "reactive" astrocytes
> - Formation of scar tissue surround areas
>   of tissue destruction

<u>Oligodendrocytes</u>

These are small cells possessing few processes, and they are responsible for myelin formation within the C.N.S. Cytoplasm does not possess gliofilaments or glycogen. Oligodendrocyte location may be perineuronal, interfascicular or perivascular.

These cells may swell as reaction to injury to the nervous system similar to the response of astrocytes

<u>Microglia</u>

These are the only cells in the C.N.S. of mesodermal origin. (All others are derived from ectoderm). They are small oval cells with numerous processes. They are capable of migration and phagocytosis (garbage collectors of C.N.S.), thus becoming "Gitter" cells.

PERIPHERAL NERVES, GANGLIA AND NERVE TERMINALS

PERIPHERAL NERVES

Most peripheral nerves are comprised of nerve fibers which are either afferent (sensory) in function of alternatively efferent (motor), and these fibers may be either well myelinated or poorly myelinated.

If a peripheral nerve is examined after being cut in cross section, the nerve fibers can be seen to be grouped in bundles which are referred to as funiculi.

The entire nerve comprising its component funiculi, is seen to be surrounded by the epineurium, which is a connective tissue sheath containing small blood vessels (vasa nervorum)

Connective tissue surrounding each funiculus in the peripheral nerve comprises the perineurium.

Within each funiculus can be seen fine strands of connective tissue surrounding individual nerve fibers comprising the funiculus, and these connective tissue strands constitute the endoneurium.

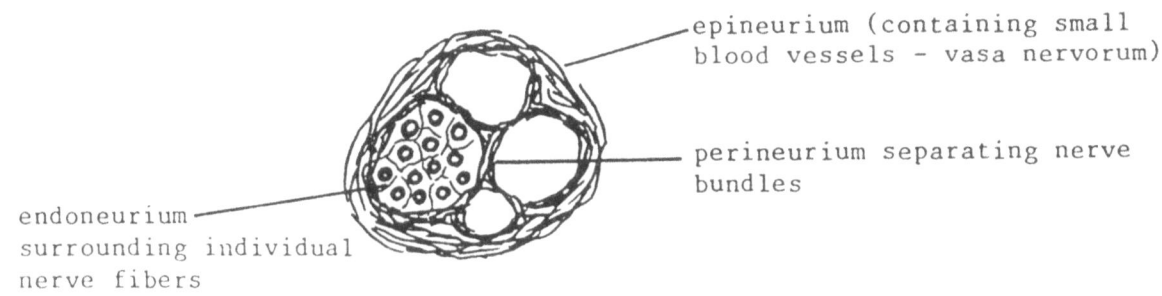

Cross Section of a Peripheral Nerve

NOTE: The amount and thickness of epineurium and perineurium may vary at different locations in a peripheral nerve. Frequently the perineurium is thicker in the more distal portions of a peripheral nerve than it is proximally. This could be of some importance in attempting surgical repair following peripheral nerve injuries.

GANGLIA

Discrete clusters or aggregates of neuron cell bodies located outside the central nervous system are called ganglia. They may be composed of either sensory neurons (e.g. dorsal root ganglia) or autonomic effector neurons (e.g. coeliac ganglion, ciliary ganglion).

Sensory Ganglia

The clusters of nerve cell bodies comprising these ganglia are surrounded by a connective tissue capsule.

This capsule, made up of collagen and elastic connective tissue, is continuous with the epineurium of the peripheral nerve. The inner portion of the capsule connective tissue fibers extends into the substance of the ganglion as fine supporting fibers, which also carry capillaries into the depths of the ganglion. These finer strands of connective tissue are continuous with endoneurium fibers of the peripheral nerve fiber.

Small flattened satellite cells may be seen in close proximity to the ganglion cell bodies.

The ganglion cell bodies (unipolar neurons) are generally located towards the periphery of the ganglion, with the central region being occupied by the cell processes extending from the cell bodies.

Ganglion cells usually possess Nissl substance which may appear as discrete clumps of Nissl substance in a clear cytoplasm. This appearance is typical of sensory ganglion cells which give rise to more heavily myelinated fibers and are associated with elaborate, more specialized nerve endings.

Other sensory ganglion cells may show a slightly darker staining cytoplasm possessing a finer, dustlike dispersed Nissl substance. These are usually cells possessing less well myelinated fibers, which are associated with less complex, simpler types of nerve endings.

ganglion cell

satellite cells

Relationship of Satellite Cells to Ganglion Cells

Autonomic Ganglia

Sympathetic chain ganglia possess a connective tissue capsule similar to that seen surrounding dorsal root ganglia. Terminal parasympathetic autonomic ganglia on the viscera tend to be supported by the connective tissue of the organ innervated.

Neurons of autonomic ganglia are of the multipolar type and are generally small in size. The cells are disposed evenly throughout the ganglion and are interspersed with nerve fibers. Nissl substance is coarser than that seen in the sensory ganglia, and binucleate cells are commonly present. Satellite cells are seen in relation to cell bodies of neurons. In general, well myelinated fibers seen in autonomic ganglia belong to preganglionic neurons, whereas poorly myelinated fibers are processes of postganglionic neurons.

INJURIES TO PERIPHERAL NERVES

Severe crush injuries, compression or entrapment, or transection (either partial or complete) may result in the following sequence of events:

1. Loss or reduction of neural function (speed of neural transmission may be reduced leading to total lack of transmission, leading to sensory loss and motor weakness).
2. Swelling, disintegration and degeneration of axons, followed by degeneration of the myelin sheaths, distal to the site of injury. Note: Myelin degeneration also proceeds proximally from the injury site for one or two myelin segments.
3. Schwann cells remain and proliferate.
4. Phagocytes migrate to the region and remove debris resulting from the degeneration of axons and myelin sheaths.
5. There is increased metabolic activity in the cell bodies of injured neurons with chromatolysis of Nissl substance.
6. Growth of axonal sprouts from the proximal stump of the injured axon.
7. Sprouts encounter "tubes" of Schwann cells distal to the injury site and regenerated axons grow along the course of Schwann cell "tubes".
8. Eventual reestablishment of distal neural connections.

Reestablishment of distal connections precisely as they were prior to injury rarely achieves perfection.

NOTE: Maximum regrowth rate under optimal circumstances is approximately 2 mm per day.

NOTE: Several disease states and various toxins may give rise to peripheral neuropathies, e.g. diabetic polyneuritis, arsenical polyneuritis, organophosphate toxicity etc.

FUNCTIONAL COMPONENTS OF PERIPHERAL NERVES

Primarily determined by the work of C. Judson Herrick

Spinal nerves generally contain fibers of four functional varieties.

These are:
                General somatic afferent (G.S.A.)
                General visceral afferent (G.V.A.)
                General visceral efferent (G.V.E)
                General somatic efferent (G.S.E.)

G.S.A. - Sensory innervation of skin, muscles, tendons, and joints
G.V.A. - Convey impulses arising from blood vessels and surrounding tissues, and visceral organs.

G.V.E. - (Autonomic) fibers supplying cardiac muscle, smooth muscle, blood vessels
      and glands.
G.S.E. - Motor fibers innervating striated muscles derived from somites.

<u>Cranial Nerves</u> may have some of the above mentioned components, but may also
        possess additional functional components not found in spinal nerves.

These are:
S.V.A. - Special visceral afferent (sensory for chemical receptors, e.g. taste,
      smell).
S.V.E. - Special visceral efferent (to muscles derived from branchial arches, e.g.
      muscles of mastication).
S.S.A. - Special somatic afferent (special senses e.g. vision, hearing).

NOTE: No one cranial nerve possesses <u>ALL</u> of the functional components.

NERVE TERMINALS

Afferent (Sensory).
      (I)   Non encapsulated ("free" endings).
     (II)   Encapsulated with coiled and branched nerve processes.
    (III)   Encapsulated with straight, unbranched nerve processes.
     (IV)   Encapsulated with other connective tissue elements
          incorporated in the capsule.

I. <u>Free Endings</u> (non encapsulated)
Myelination terminates prior to nerve ending.
Axon may branch profusely (e.g., visceral afferents).
May end in squamous epithelium of skin (pain fibers).
May end as cup-like ending in contact with a particular epithelial cell (Merkel's
tactile disc).
Also coiled endings surrounding hair follicles. (Peritrichial endings).

Nerve Endings in Skin

II. <u>Thinly encapsulated endings</u> (e.g., Meissner's corpuscles)
Found in deep epidermis particularly in certain regions. Fingers, palm of hand,
toes and sole of foot, lips, nipple, clitoris and penis.

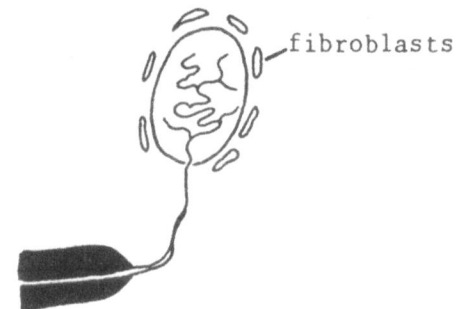

fibroblasts

Axon is coiled and branched inside the thin
capsule (said to respond to forces
perpendicular to skin surface as contrasted
with Merkel's discs which supposedly respond
to forces tangential to skin surface).
Discriminatory tactile stimuli.

Meissner's Corpuscle

Others of this structural type

End bulbs of Krausse are spherical
and larger than Meissner's corpuscles.
Possibly stimulated by cold temperatures.

End Bulb of Krausse

Golgi-Mazzoni Corpuscles are smaller
than Krausse Bulbs. They possess a
slightly thicker capsule and the
terminals are not as well coiled.
Said to be responsive to pressure.

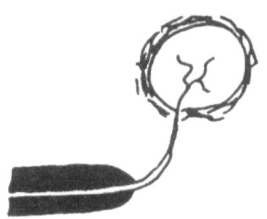

Golgi-Mazzoni Corpuscle

III. <u>Encapsulated endings with thick capsules</u>, e.g. Pacinian corpuscles.

Thick, laminated connective tissue capsule with
straight nerve terminals within. Layers of collagen
fibers form lamellae around the nerve ending in the
fashion of an onion skin. May be very large in size
(up to 3 or 4 mm). These are deep pressure
receptors found in subcutaneous connective tissue,
adventitia of blood vessels, and around joint
capsules.

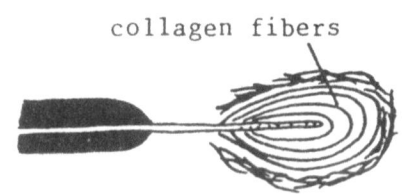

collagen fibers

Pacinian Corpuscle

IV. <u>Terminations containing other tissue elements with the capsule.</u>
e.g. Muscle spindles and tendon spindles. All contain tissue elements (such as muscle or tendon) as well as nerve terminals within the capsule.

Muscle Spindles _ (Muscle stretch receptors)

These are found in striated muscle, where they are surrounded by a connective tissue capsule. Most numerous in muscles of the extremities. Least in extraocular eye muscles.

May vary from 0.75 mm to 10 mm in length (Average length 2-4 mm). The capsule contains modified striated muscle fibers (intrafusal fibers). These fibers number from 2-10 per capsule and are striated except at midpoint (equator of spindle). Nuclei of intrafusal fibers are accumulated at the equator of the spindle (sometimes termed nuclear bag). The ends of the intrafusal fibers receive motor innervation (gamma efferent fibers).

Muscle spindles possess sensory innervation consisting of two types:

1. <u>Annulospiral endings</u>
- coils from terminal of nerve fiber extending in both directions along the length of intrafusal fibers. Thick, primary sensory endings.

2. <u>Flower-spray endings</u>
- branched, knobbed terminals located at the ends of intrafusal fibers. Thinner, secondary sensory endings.

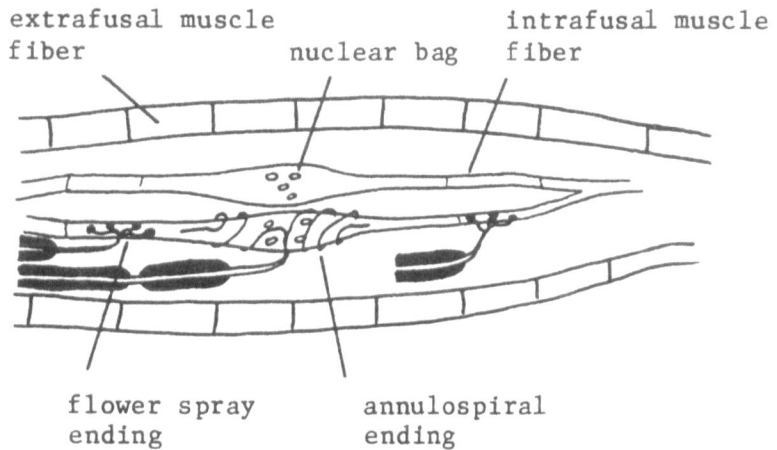

Muscle Spindle in Striated Muscle

<u>Efferent Terminations</u>

<u>Motor end plates</u> found on striated muscle fibers are formed by expansion of the terminal end of the axon.

Myelinated axons may branch at a node of Ranvier to innervate several muscle fibers additional to those innervated by the main terminal axon.

NOTE: One anterior horn cell and all the muscle fibers supplied
            by it constitute a physiological motor unit.

Larger back muscles - one neuron may innervate up to 100 muscle fibers.
Small intrinsic hand muscles - one neuron may only innervate 2 or 3 muscle fibers.
Fibers lose their myelin prior to termination.

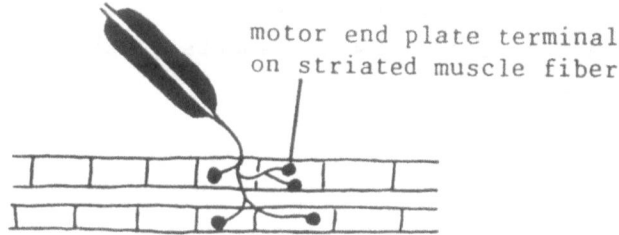

motor end plate terminal
on striated muscle fiber

Motor Nerve Terminals

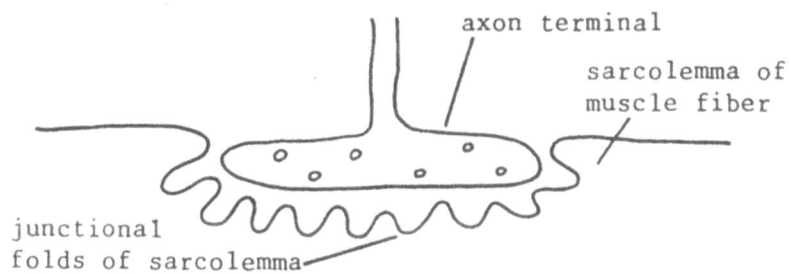

axon terminal

sarcolemma of
muscle fiber

junctional
folds of sarcolemma

Enlarged View of the Neuromuscular Junction

NOTE: Disorders of the neuromuscular junction may occur in the
          presence of normal nerve conduction and normally functioning
          muscle fibers. For example, myesthenia gravis - weakness
          and rapid fatigue of muscles thought to be due to
          insufficient production of acetylcholine, or lack of
          available receptor sites at the muscle membrane.
          Administration of edrophonium chloride (Tensilon) results in
          dramatic improvement for a short time.

Endings on Smooth and cardiac muscle, and acini of glands.

The manner of termination may vary somewhat.
Some processes terminate as swellings or knobs on muscle fibers.
Other fibers may form a terminal network between muscle fibers or glandular
secretory cells.

GENERAL SENSORY SYSTEMS

Sensory systems transmit information arising from receptor structures at various
bodily locations into the central nervous system. General sensation refers to such
things as pain, thermal sensation, touch and pressure etc. Other sensory systems
are concerned with various special senses, such as vision, olfaction and hearing.
Special senses are, of course, part of the function of some of the cranial nerves.

General sensation arising from the neck, trunk and limbs travels via sensory fibers
contained in spinal nerves, and enters the central nervous system at the appropriate
spinal cord level. General sensations arising from structures in the head are
obviously carried in the appropriate cranial nerves, and therefore enter the central
nervous system at the level of the brainstem.

CHARACTERISTICS OF SENSORY SYSTEMS

Neurons possessing peripheral processes which terminate as sensory receptor
structures are generally located in sensory ganglia (e.g. dorsal root ganglia) in
the company of other such neurons. The central process of these primary sensory
neurons enters the central nervous system where it usually branches to some extent
before synapsing on other neurons.

Synaptic connections made centrally may result in a variety of possibilities
regarding the subsequent course taken by incoming impulses. Some of these
possibilities are summarized below.

1. Ascending sensory pathways traveling from the level of entry up through higher
   levels of the nervous system, to transmit sensation to the cerebral cortex where
   sensation may be recognized at a conscious level.

2. Reflex connections which effect synaptic contact with motor neurons located
   either at the level of entry or at some more remote location within the nervous
   system.

3. Pathways which project information into the cerebellum.

4. Diffuse pathways which provide sensory input into the reticular formation.

REFLEX CONNECTIONS

These may be relatively simple, where the sensory neuron process synapses directly
upon a motor neuron (monosynaptic reflex). Others may influence motor neuron
activity indirectly via one or more intermediate neurons interposed between the
sensory neuron and the motor neuron. These multisynaptic reflex connections are
more common than are monosynaptic connections.

At spinal cord levels, reflex motor response to sensory input may occur at the same
spinal level as the stimulus (intrasegmental reflex arc), at a different spinal
segmental level (intersegmental reflex arc). Similarly, reflex motor response may
occur on the same side of the body as the side of the stimulus (ipsilateral reflex),
or alternatively may occur on the side of the body opposite from the side of the

stimulus (contralateral reflex).

SENSORY PATHWAYS TO CONSCIOUS LEVELS

Ascending in the spinal cord to higher centers in the nervous system, are two particularly important pathways:

  I. THE SPINOTHALAMIC SYSTEM - pain, thermal sensation and
                        crude touch.
  II. THE POSTERIOR COLUMN SYSTEM - vibratory sense, two point
                        discrimination, joint position
                        sense and discriminatory touch.

RECEPTORS

Naked nerve endings (pain).
Peritrichial endings (pain and crude touch).   } . . . . Spinothalamic system
Merkel's tactile discs (crude touch).
Krausse end bulbs (thermal receptors).

Pacinian corpuscles (vibration and pressure).
Golgi-Mazzoni corpuscles (pressure).   } . . Posterior column system
Flower-spray endings (limb and joint position).
Annulospiral endings (limb and joint position).

Primary neuron cell bodies are located in dorsal root ganglia and their peripheral processes terminate in relation to sensory receptors. The larger neurons possessing well myelinated axons are concerned with the posterior column system (joint position, etc.). Smaller cells with less heavily myelinated axons are concerned with the spinothalamic system (pain etc.)

Central processes constitute the dorsal root of the spinal nerves. At the attachment of the dorsal roots to the spinal cord, the larger fibers (vibratory etc.) are more medially located than the pain fibers.

COURSE TAKEN WITHIN THE C.N.S.

I. SPINOTHALAMIC SYSTEM

Fibers enter the dorsolateral fasciculus (also known as the zone of Lissauer). They bifurcate and take a short ascending and descending course (approximately 1 or 2 spinal segments).

They then synapse at various locations on cells in the dorsal horn grey matter. Axons of these secondary neurons located in the dorsal horn, cross to the contralateral side of the cord via the ventral white commissure, to form the spino-thalamic tracts (lateral and ventral), and the spinotectal tract.

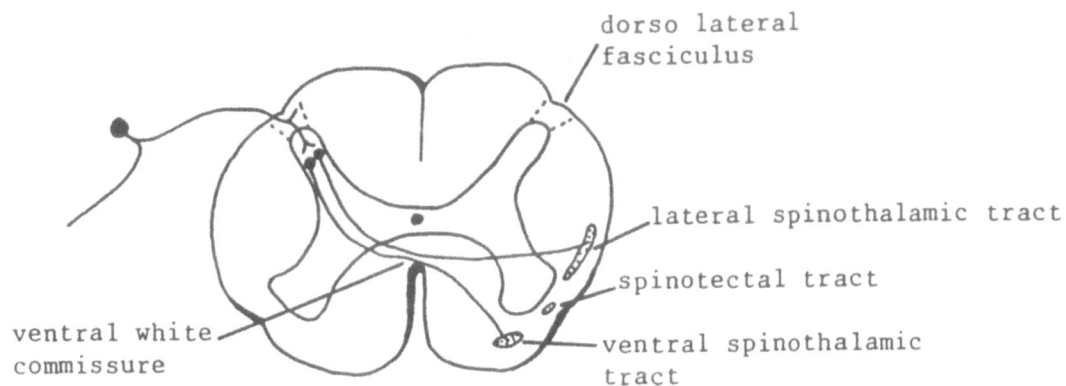

Position of the Spinothalamic System in the Spinal Cord

Lateral spinothalamic tract - located in the anterolateral portion of the lateral funiculus and conveys pain and thermal sense (from the contralateral side of body). This tract is laminated so that fibers from lower cord levels are located dorsolaterally. Fibers entering from higher levels are located ventromedially.

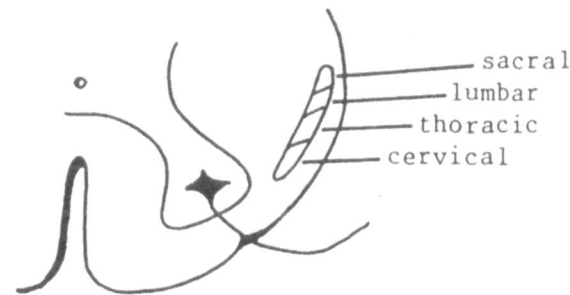

Lateral Spinothalamic Tract in Upper Cervical Cord

Ventral spinothalamic tract - found in the ventral funiculus and conveys sensations of crude touch. This tract is not as well laminated as the lateral spinothalamic tract.

Both these tracts ascend through the spinal cord up to the lower medulla, where they fuse into a single tract - the spinal lemniscus, at the level of the inferior olivary nucleus. The spinal lemniscus terminates in the posterior ventral lateral (P.V.L.) thalamic nucleus.

Spinotectal fibers, which occupy an intermediate position between the spinothalamic tracts, terminate in the midbrain tectum mediating reflex movements of the head etc. in response to pain.

Tertiary neurons in the P.V.L. nucleus send axons via the thalamic radiations in the posterior limb of the internal capsule, to terminate primarily in the general sensory cortex of the cerebral hemisphere (postcentral gyrus).

NOTE: In the neurological examination, use of light pin prick
         (sharp versus dull) and hot and cold test objects on the
         skin surface tests the integrity of the spinothalamic
         system.

Spinoreticular System

Other fibers mediating pain sensation arise from the intermediate spinal cord grey
matter and ascend on both sides of the cord in the lateral and ventral funiculi.
These constitute the rather diffuse spinoreticular tracts which synapse upon the
brainstem reticular formation neurons. From here, reticulothalamic fibers ascend to
the intralaminar and centromedian thalamic nuclei, as well as the posterior ventral
lateral thalamic nucleus.

II. POSTERIOR COLUMN SYSTEM

Fibers enter the spinal cord at the more medial portion of dorsal roots and
immediately become a part of the posterior white funiculus of the cord, where they
ascend the entire length of the cord without interruption. Fibers from lower cord
segments are located adjacent to the midline (fasciculus gracilis), higher segments
located more laterally in the posterior funiculus (fasciculus cuneatus).

Posterior column fibers terminate on cells of the respective relay nuclei (nucleus
gracilis and nucleus cuneatus) located in the lower medulla.

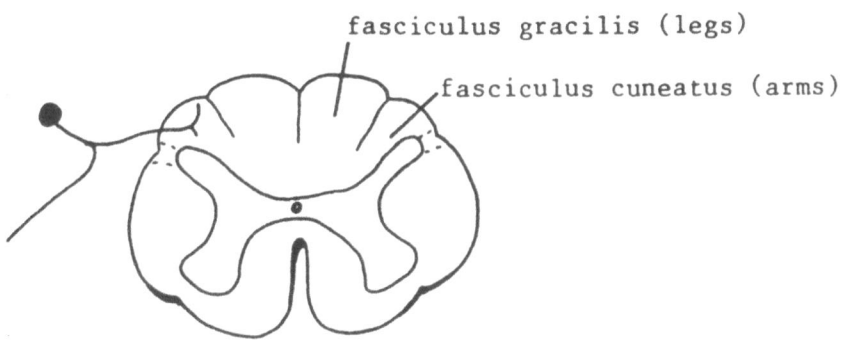

Spinal Cord showing Posterior Columns

Secondary fibers originating from these nuclei take a ventromedial course across the
midline (as the internal arcuate fibers) to form the medial lemniscus on the
contralateral side of the brainstem, adjacent to the midline.

The medial lemniscus courses through the medulla, becomes horizontally situated at
the level of the pons, and eventually terminates in the posterior ventral lateral
thalamic nucleus. Tertiary fibers are projected to the postcentral gyrus via the
posterior limb of the internal capsule.

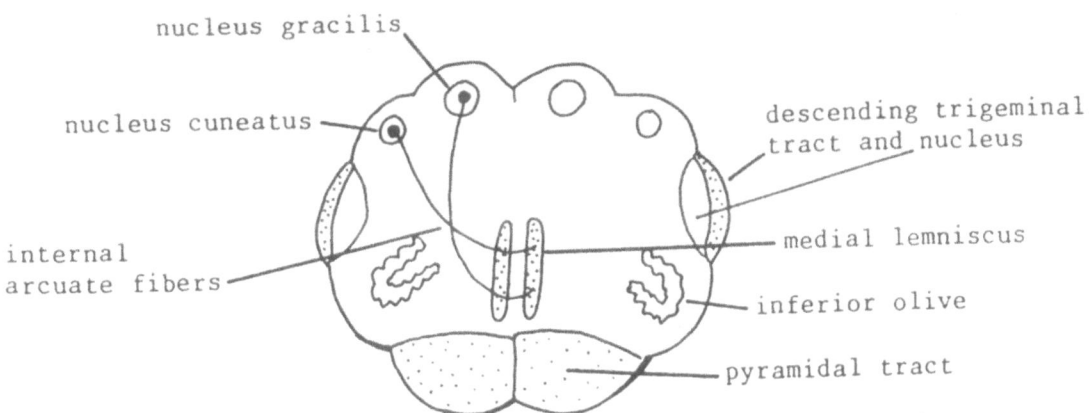

Cross Section of the Lower Medulla

NOTE: Unilateral lesions in the brainstem which may simultaneously
          involve both medial lemniscus and spinothalamic fibers,
          result in contralateral sensory deficits for posterior
          column sensation, and also for pain and thermal sensation.
          (i.e. both types of sensory deficit will be on the same side
          of the body)

          However, unilateral cord lesions affecting both systems
          simultaneously such as hemisection or hemicompression (Brown
          Sequard syndrome) result in contralateral loss of pain and
          thermal sense, and ipsilateral loss of vibratory sense, etc.

NOTE: In neurological testing, diminished posterior column function-
          ing may be revealed by the patient's inability to perceive
          vibrations of a tuning fork, and inability to determine the
          position of toes or fingers with closed eyes. For example:
          In pernicious anemia and tabes dorsalis.

THE SPINOCEREBELLAR SYSTEM

Sensory information from musculoskeletal receptors (muscle spindles, joint
receptors) is relayed into the cerebellum to aid in cerebellar coordination of
muscular activity. For this purpose, some posterior column fibers synapse on cells
of the nucleus dorsalis (Clarke's column) located in the ipsilateral lumbar and
lower thoracic cord. From here, fibers project to the cerebellum as the spinocerebellar tracts (dorsal and ventral).

The dorsal spinocerebellar tract enters the cerebellum as part of the restiform body
(inferior cerebellar peduncle). The ventral spinocerebellar tract ascends to the
level of the lower midbrain and then travels via the brachium conjunctivum (superior
cerebellar peduncle) to enter the cerebellum.

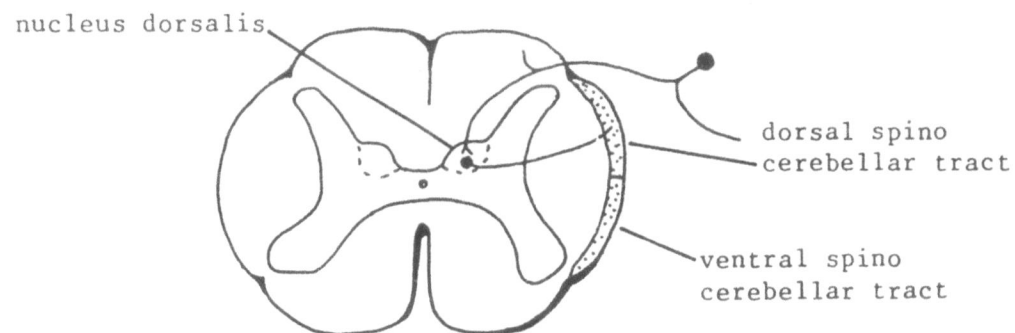

nucleus dorsalis

dorsal spino
cerebellar tract

ventral spino
cerebellar tract

**Cross Section of the Spinal Cord
in the Lumbar Region**

NOTE: Spinocerebellar fibers ascend on the same side of the cord
as the side of origin, except for a small number of fibers
in the ventral spinocerebellar tract arising from the
contralateral nucleus dorsalis

NOTE: There is no nucleus dorsalis located in the cervical cord.
Impulses from cervical levels destined to reach the
cerebellum ascend in the fasciculus cuneatus to the lower
medulla, and terminate in the accessory cuneate nucleus,
located just lateral to the main cuneate nucleus. Neurons
of the accessory cuneate nucleus perform the same function
as nucleus dorsalis neurons, i.e., relay into the cerebellum. Cerebellar projection is via the direct arcuate
fibers, which enter the cerebellum as part of the ipsilateral restiform body.

TRIGEMINAL SYSTEM AND TASTE

THE TRIGEMINAL NERVE (CRANIAL N.V)

This cranial nerve is the principal cranial nerve responsible for transmitting general sensation arising from many structures in the head.

The cell bodies of the sensory unipolar neurons of this cranial nerve reside in the trigeminal ganglia (also called the Gasserian ganglion or the semilunar ganglion). The ganglion is located in Meckel's cave (a shallow depression located on the medial portion of the anterior slope of the petrous bone in the middle cranial fossa). From this ganglion, the peripheral processes of the ganglion cells give rise to three major divisions of the trigeminal nerve, which are distributed over various regions of the head.

The first division is the Ophthalmic, which provides sensory innervation to skin of the forehead and anterior scalp (as far as the vertex of the skull), upper eyelid, cornea and conjunctiva of the eye, skin on the dorsum of the nose, mucous membranes of the nasal vestibule and the frontal sinuses.

The second division is named the Maxillary, whose branches innervate skin of the upper cheek and lateral portion of the nose, skin of the anterior portion of the temple, the upper lip, the mucous membrane of the roof of the mouth and the teeth of the upper jaw.

The Mandibular, or third division, innervates skin of the lower lip and lower jaw, a portion of the cheek and temple, the external ear, the mucous membrane of the lower portion of the mouth, the floor of the mouth, the tongue and teeth of the lower jaw.

NOTE: All three divisions, in addition to the above, give rise to small branches which innervate the meninges residing in the supratentorial compartment of the cranial cavity.

MOTOR DIVISION OF TRIGEMINAL, S.V.E.

Although the trigeminal nerve is largely sensory in nature, it does, in addition, contain a motor portion. The origin of these motor fibers lies in the trigeminal motor nucleus located in the pons which is the source of the motor fibers innervating the muscles of mastication (temporalis, masseter, medial pterygoid and lateral pterygoid), and also the tensor tympani and tensor veli palatini muscles.

THE TRIGEMINAL SENSORY PATHWAYS

The central processes of all the nerves in the trigeminal ganglion contribute the sensory root of the trigeminal nerve, which travels across the floor of the cranial cavity to attach to the brainstem at the level of the pons. These fibers gain access to the pons by penetrating at right angles through the fibers of the brachium pontis. Once within the dorsolateral portion of the pons, these sensory fibers form a short ascending trigeminal tract terminating in the main sensory trigeminal nucleus, and a longer descending tract terminating in the descending (or spinal) trigeminal nucleus. (NOTE: The descending tract and nucleus extends down throughout the caudal pons and the extent of the medulla into the spinal cord as low as the 3rd cervical spinal cord level. The main sensory nucleus is located in the rostral pons.)

The three divisions of the trigeminal are represented in the descending tract and nucleus with the mandibular division most dorsal, ophthalmic division most ventral, and the maxillary division in an intermediate position.

In the upper cervical cord, down to C.3, the neurons of the dorsal horn grey matter are intermingled with the neurons of the most caudal portion of the descending trigeminal nucleus. NOTE: Sensory fibers of the 2nd and 3rd cervical spinal nerves innervate skin at the back of the scalp as far forward as the skull vertex.

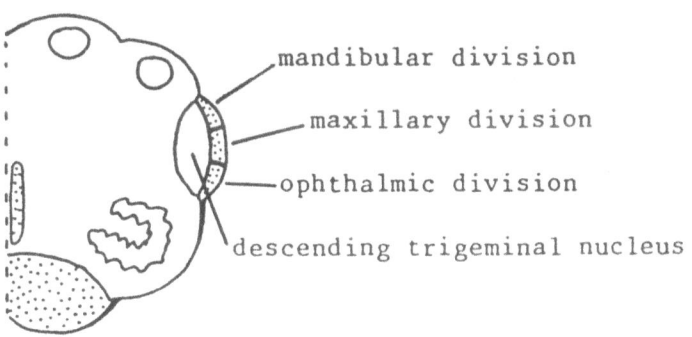

Descending Trigeminal Tract in the Medulla

NOTE: A small area of skin around the external ear has innervation
            for general sensation via sensory fibers of cranial nerves
            VII (geniculate ganglion), IX (superior ganglion), and X
            (jugular or superior ganglion). The central connections of
            these sensory fibers are the same as the central trigeminal
            sensory fibers.

Discriminatory touch, two point discrimination, etc.: handled by the main sensory nucleus.

Crude touch: Processed by the rostral portion of the descending nucleus.

Pain and Thermal Sense: Relayed through the caudal portion (caudal to the inferior olivary nucleus) of the descending trigeminal nucleus.

NOTE: The mesencephalic nucleus of the trigeminal system, located
            in the periaqueductal grey matter of the midbrain, is
            composed of unipolar neurons. These are similar to the
            neurons of the trigeminal ganglion. Their single processes
            divide into a peripheral and central process whilst still
            within the brainstem, and these processes constitute the
            mesencephalic tract of the trigeminal nerve. The peripheral
            processes exit the brainstem and terminate upon the muscle

spindles in the muscles of mastication, which receive their
motor innervation via the motor division of the trigeminal
nerve. The central processes of the cells synapse upon
cells in the main sensory nucleus, and also upon the cells
of the motor nucleus of the trigeminal, located medial to
the main sensory nucleus.

Secondary neurons from the main sensory nucleus and the descending trigeminal
nucleus project axons across the midline to form the <u>Trigeminal Lemniscus</u>, which is
the main trigeminothalamic pathway. This travels in close association with fibers
of the medial lemniscus, to terminate in the <u>Posterior Ventral Medial</u> thalamic
nucleus.

Some secondary fibers arising from the main sensory nucleus ascend as an uncrossed
tract, the dorsal trigeminothalamic tract, terminating in the Posterior Ventral
Medial thalamic nucleus.

Tertiary fibers from this nucleus travel via the internal capsule to terminate in
the inferior portion of the postcentral gyrus. Additional connections of cells of
the main sensory and descending trigeminal nuclei are into the brainstem reticular
formation and into the cerebellum.

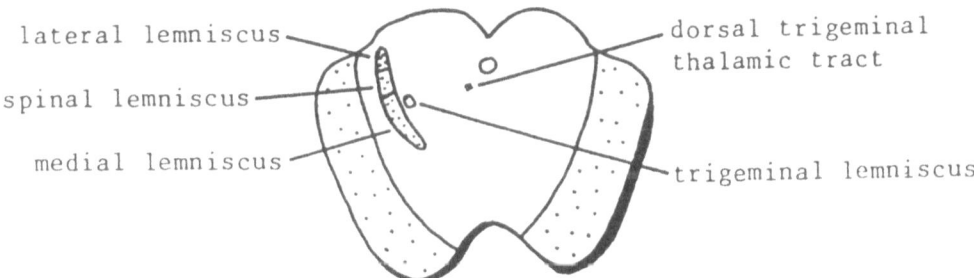

Midbrain Section showing the Lemniscal Systems

A number of reflexes involve the trigeminal nerve as part of the reflex arc.
For example:

        The corneal reflex (trigeminal afferent limb, facial-motor limb)
        The lacrimal reflex
            (trigeminal - sensory, facial nerve parasympathetics - motor)
        The jaw jerk reflex - Stretch reflex of the masseter muscle
                    (trigeminal sensory - afferent limb,
                      trigeminal motor - efferent limb)

NOTE: Trigeminal neuralgia - an excruciatingly painful condition,
           the etiology of which is unclear, which may affect one or
           more of the peripheral divisions of the trigeminal nerve.
           May be triggered by extremely light stimulation of the skin
           surface of the affected portion of the nerve.

TASTE

Taste buds in the anterior 2/3 of the tongue are innervated by sensory fibers in the facial nerve (cell bodies located in the geniculate ganglion). The posterior 1/3 of the tongue taste buds are innervated by fibers of the glossopharyngeal nerve from the petrosal (or inferior) ganglion (IX). Fibers of the vagus nerve innervate taste receptors on the epiglottis from the nodose (inferior) ganglion (X).

Centrally, taste fibers form the solitary tract and terminate on the cells of the solitary nucleus which surrounds it.

Secondary fibers from the solitary nucleus ascend as part of the contralateral medial lemniscus, terminating in the posterior ventral medial thalamic nucleus. Others make connections with the brainstem reticular formation and the salivatory nuclei (superior and inferior) in the medulla which provide innervation to the salivary glands. The superior salivatory nucleus is the origin of parasympathetic preganglionic fibers of the facial nerve. The inferior salivatory nucleus provides parasympathetic preganglionics of the glossopharyngeal nerve.

Cortical regions concerned with taste are the opercular portion of the post central gyrus and the insula cortex.

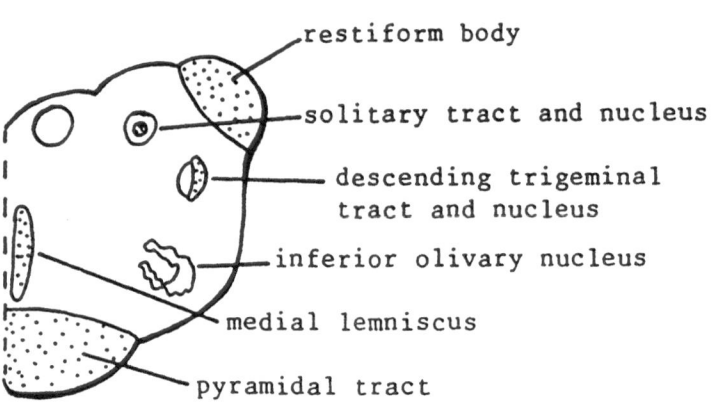

restiform body

solitary tract and nucleus

descending trigeminal tract and nucleus

inferior olivary nucleus

medial lemniscus

pyramidal tract

Position of the Solitary Tract in the Medulla

MOTOR SYSTEMS - LOWER MOTOR NEURONS AND PYRAMIDAL SYSTEM

LOWER MOTOR NEURONS

Central nervous system system influence is exerted on muscles in the body by the activity of motor neurons located within the C.N.S., the axons of which form the motor fibers of peripheral nerves. Action potentials over those motor axons result in release of neurotransmitter substance at the motor end plate and hence muscle contraction.

The term motor unit may be used to describe such a motor neuron, its peripheral axon, and all the muscle fibers which it innervates.

The term lower motor neuron is also used to refer to the motor neuron and its axon, excluding the muscle fibers innervated. Thus, examples of lower motor neurons would be the motor neurons located in the anterior horn of spinal cord gray matter, and also neurons residing in the brainstem from which arise motor fibers innervating motor structures in the head (e.g. the hypoglossal nucleus).

Lower motor neuron activity is modified in a number of ways, such as reflex connections, reticular formation influence, vestibular activity, cerebellar influence, basal ganglia and cerebral cortex.

Although these central connections have a great deal of control over the overall functioning of lower motor neurons, and hence peripheral muscle activity, absence or nonfunctioning of lower motor neurons makes C.N.S. control of muscles impossible.

Lower motor neuron malfunction, whether due to external damage or disease, will result in inability to control peripheral musculature.

Disorders of lower motor neurons result in muscle weakness (or complete motor paralysis if sufficient motor units are affected), reduction in muscle tone, diminished muscle stretch reflexes and flaccidity of the affected musculature.

The early signs of lower motor neuron disorder may be evidenced as fasciculations or fibrillations of the affected muscles. Fasciculation can be visually observed as random muscle contractions of the muscles in question, fibrillations are much smaller contractions which may only be detected by the use of electromyography (E.M.G.).

PYRAMIDAL SYSTEM

This important mechanism which influences lower motor neuron activity, is sometimes called the upper motor neuron system. It is comprised of cells and fibers arising from the cerebral cortex which project without synapse to lower motor neurons and adjacent reticular formation in the brainstem and spinal cord.

The corticospinal part of the pyramidal system extends into the spinal cord regulating lower motor neurons in spinal nerves.
The corticobulbar portion only extends down as far as cranial nerve motor nuclei.

Functionally this system is concerned with the initiation of discrete, finely controlled voluntary movement, particularly of distal musculature of the extremities and muscles involved in speech or vocalization.

ORIGIN

Pyramidal cells of a wide range of sizes, located in layers III and V of the precentral gyrus (Area 4) contribute most of the fibers to the pyramidal tract. Similar cells in the caudal portion of the superior , middle and inferior frontal gyri (Area 6), and the anterior portions of the parietal cortex contribute a few fibers.

Giant pyramidal cells found in the precentral gyrus and paracentral lobule give rise to the largest, most well myelinated fibers in the pyramidal tract. These are Betz cells, 75% of which are located in the region which controls the lower extremity. However, these cells only provide about 3% of the total number of fibers in this system. The remainder of the pyramidal tract fibers arise from the smaller pyramidal cells not included in the term Betz cells.

Cortical Representation of Body Parts

Like a manikin draped over the cerebral convexity with the feet adjacent to the paracentral lobule in the interhemispheric fissure. The head region resides at the inferior part of the motor cortex next to the Sylvian fissure. The upper extremity is represented in an intermediate position, between face and legs.

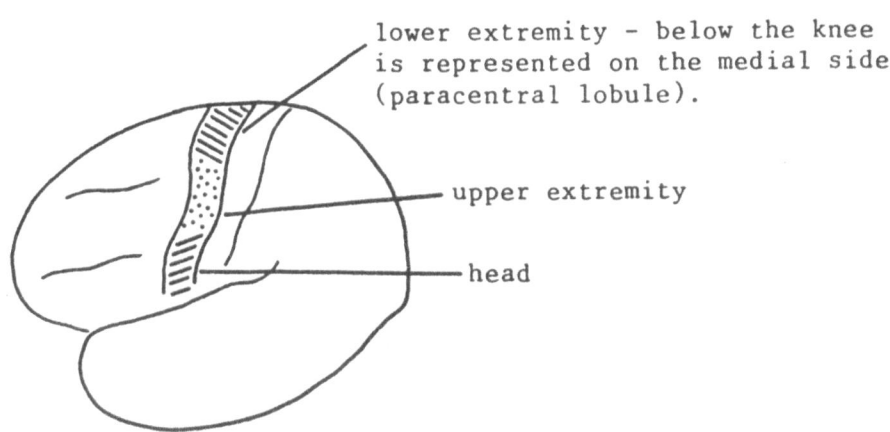

lower extremity - below the knee
is represented on the medial side
(paracentral lobule).

upper extremity

head

Representation of the Body Parts on the
Motor Cortex

CORTICOSPINAL FIBERS

These descend down the neuraxis via the internal capsule and the cerebral peduncle, on the same side as their origin as far as the caudal medulla.

At the medullary/cord junction, approximately 90% of the corticospinal fibers cross over to assume a position in the contralateral side of the spinal cord (decussation of the pyramidal tract).

Crossed fibers form the lateral corticospinal tract (in the posterior part of the lateral funiculus). The remainder of the fibers which do not cross at the medullary decussation (10%) descend as the ventral corticospinal tract, in the ventral funiculus.

These fibers however, do eventually cross the midline at the level of their termination, and so, like the others, finally terminate on the side of the neuraxis contralateral to the side of their origin in the cerebral cortex.

<u>Course of Pyramidal Tract Fibers</u>

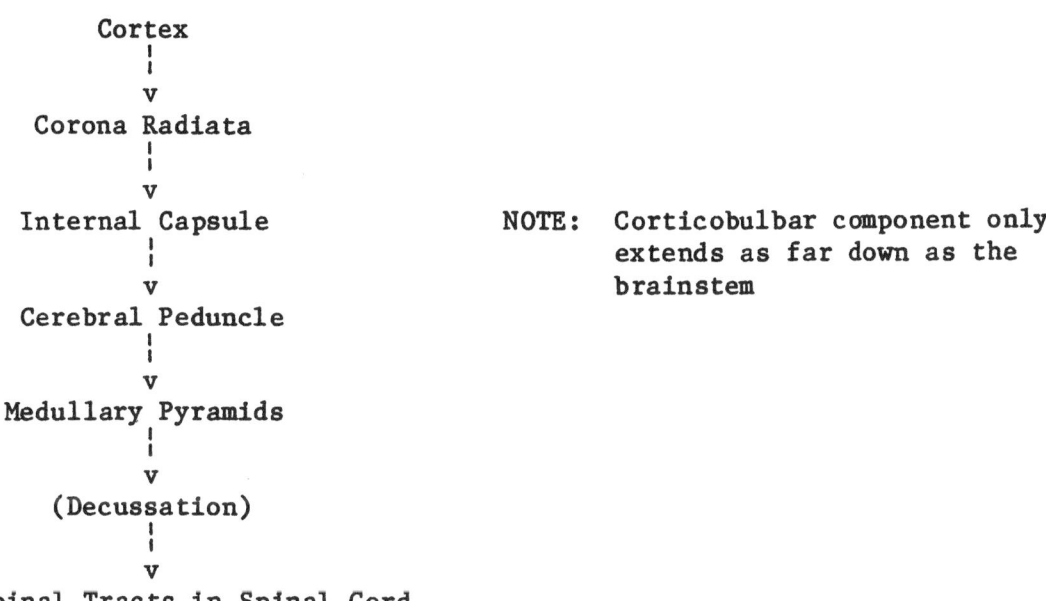

```
                    Cortex
                      ¦
                      v
                Corona Radiata
                      ¦
                      v                 NOTE:  Corticobulbar component only
              Internal Capsule                 extends as far down as the
                      ¦                         brainstem
                      v
             Cerebral Peduncle
                      ¦
                      v
            Medullary Pyramids
                      ¦
                      v
                (Decussation)
                      ¦
                      v
Corticospinal Tracts in Spinal Cord
```

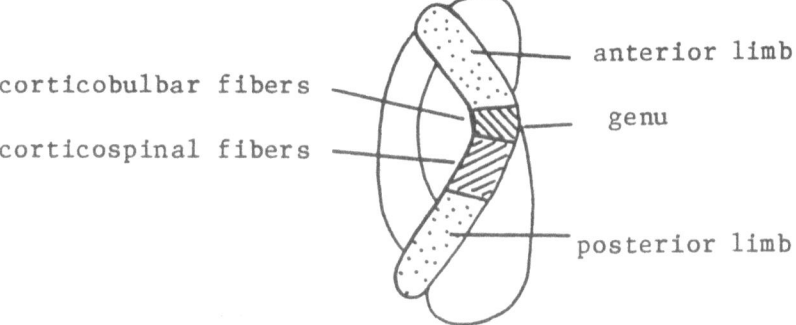

Horizontal Section of the Internal Capsule

Termination of the corticospinal tract fibers is upon anterior horn motor cells, and also upon the cells in the intermediate gray matter of the spinal cord which, in turn, synapse upon anterior horn cells. Of the pyramidal tract fibers, approximately 55% terminate in the cervical cord, 20% in the thoracic, and 25%

terminate at lumbosacral levels.

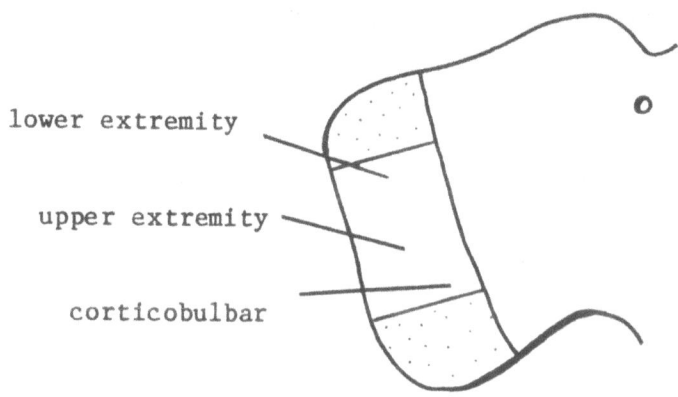

**Position of the Pyramidal Tract Fibers
in the Cerebral Peduncle**

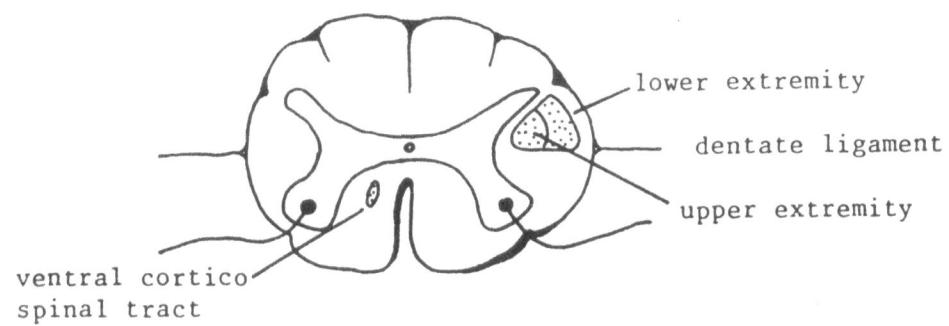

Position of the Corticospinal Tracts in the Spinal Cord

Lateral Corticospinal Tract
─────────────────────────

Fibers influencing the upper extremity are located most medially. and fibers for the lower extremity are found in the lateral portion of this tract.

Ventral Corticospinal Tract
──────────────────────────

This small portion of the corticospinal system only descends only as far as the cervical enlargement.

CORTICOBULBAR FIBERS

These pyramidal system fibers provide upper motor neuron innervation in a manner similar to the corticospinal fibers, to the motor nuclei of the brainstem. Their termination is primarily upon the nuclei of cranial nerves V (Trigeminal), VII (Facial), IX (Glossopharyngeal), XI (Spinal accessory) and XII (Hypoglossal).

The main difference between corticobulbar and corticospinal systems:

> Lower motor neurons in the spinal cord (for the extremities) receive corticospinal fibers <u>only</u> from the contralateral cortex.

> Lower motor neurons in the brainstem receive corticobulbar innervation from both contralateral and ipsilateral cerebral cortex.

<u>One Important Exception</u>

Regarding the motor nucleus of VII nerve (facial).

This nucleus comprises motor neurons of two groups. One group gives rise to axons which innervate facial muscles of expression below the level of the eye. The other group of neurons give fibers which innervate muscles above the level of the eye.

That group of facial nerve neurons innervating the lower facial muscles receives corticobulbar fibers from the contralateral cerebral cortex only (i.e. similar to lower motor neurons in the spinal cord).

The remainder of the facial nucleus (innervating facial muscles above the level of the eye) receive corticobulbar fibers from <u>both</u> contralateral and ipsilateral cerebral cortex in the same manner as other cranial nerve motor nuclei..

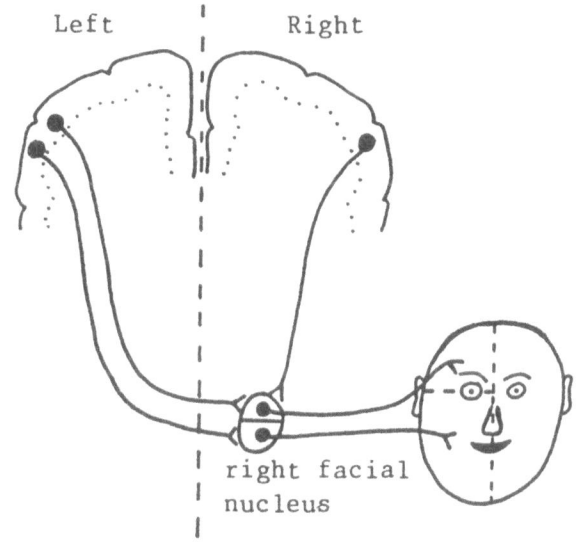

Upper Motor Neuron Innervation of the
Facial Nucleus

Finally, pyramidal tract fibers usually terminate on neurons immediately adjacent to the lower motor neurons (reticular formation neurons in the brainstem and intermediate gray matter neurons in the spinal cord). Only a small number actually synapse directly on the lower motor neurons themselves.

NOTE: Some fibers of the pyramidal system terminate in relation to
         sensory relay nuclei, e.g., gracile and cuneus nuclei, and
         trigeminal sensory nuclei.

NOTE: Interruption of the pyramidal tract at any point results in
         impaired motor function ranging from reduced muscle strength
         to dense paralysis, depending on the extent of the damage.
         Lesions above the level of decussation give rise to signs on
         the contralateral side extremities. Lesions of the lateral
         corticospinal tract in the cord result in ipsilateral signs.

Due to the compact nature of the pyramidal tract as the fibers traverse the internal
capsule, relatively small infarcts in this region can result in profound
hemiparesis, involving both extremities on the contralateral side, and also (if
involving corticobulbar fibers in the genu) contralateral facial muscle paralysis
below eye level.

NOTE: The difference between a facial nerve lower motor neuron
         lesion and an upper motor neuron lesion affecting the facial
         nerve is that in the former, all the facial musculature on
         the affected side is impaired (above and below the eye),
         whereas in the upper motor neuron lesion, impairment is to
         the contralateral lower facial musculature only with sparing
         of muscles above the eye.

         If an individual has a hemifacial paralysis with mouth
         sagging on one side, but despite this can raise both
         eyebrows or frown symmetrically, then most likely this is
         due to an upper motor neuron lesion, not a lower motor
         neuron disorder.

NEUROLOGICAL SIGNS OF PYRAMIDAL TRACT LESIONS:
(Upper Motor Neuron Signs)

         Hypertonia
         Hyperreflexia
         Clonus (due to hyperactive muscle stretch reflexes)
         Babinski sign (upgoing big toe and occasionally flaring of
                        other toes in response to stimulation of the
                        sole of the foot).

NOTE: The Babinski response can usually be elicited in very young
         children up to the age of 6 to 9 months and is a normal
         response for that age group. Disappearance of the response
         usually occurs prior to age at which the child begins to
         walk, although it may persist for some time afterwards.

THE CORTICO-PONTO-CEREBELLAR PATHWAY

At the time when the pyramidal system is actively influencing lower motor neurons,

there is simultaneous transmission of this information into the cerebellum.
Cortical discharge from widespread areas is carried via the corticopontine fibers
found in the internal capsule. These fibers continue as a portion of the cerebral
peduncle and eventually terminate as synapses upon groups of nuclei located in the
basal pons. These pontine nuclei neurons give rise to pontocerebellar fibers which
cross the midline and constitute the contralateral brachium pontis. Thus the
cerebellum is informed on a moment to moment basis about what is occurring in this
part of the motor control system.

MOTOR SYSTEMS - THE BASAL GANGLIA SYSTEM

This important portion of the C.N.S. involved in the over all regulation of motor function comprises a widely spread system of nuclei and fiber tracts, cerebral cortex and numerous sub-cortical regions. In contrast to the pyramidal system, this complex motor system influences lower motor neurons via numerous neurons and synapses, involving complex feedback mechanisms. Working in concert with the pyramidal motor system, the basal ganglia apparatus assists in the control of purposive movement. This system has a greater influence over proximal musculature than does the pyramidal system, and conversely, the pyramidal system has a greater influence upon the distal musculature.

NOTE: In addition it may be said that the basal ganglia system regulates associative movements which generally occur during the performance of normal activity. An example of the type of associative movements would be swinging the arms when walking, and changes in facial expression during conversation. These supportive motor activities may be altered or absent in the presence of basal ganglia system malfunction.

Fibers arising from cells in widespread areas of frontal and parietal cortex project onto the cells which comprise the basal ganglia. These are large nuclear groups found deep in the substance of the cerebral hemispheres.

Definitions

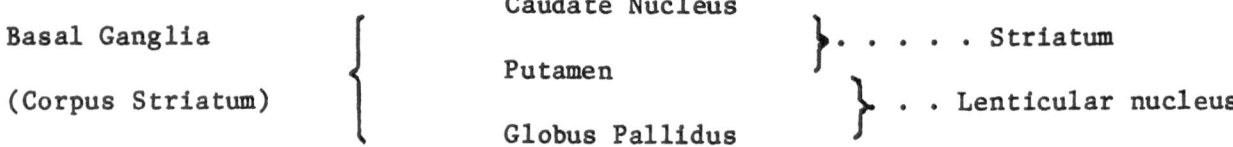

Basal Ganglia

(Corpus Striatum)

Caudate Nucleus

Putamen . . . . . Striatum

Globus Pallidus . . Lenticular nucleus

NOTE: The claustrum and the amygdaloid nucleus are also sometimes included in the general term basal ganglia, by some authors.

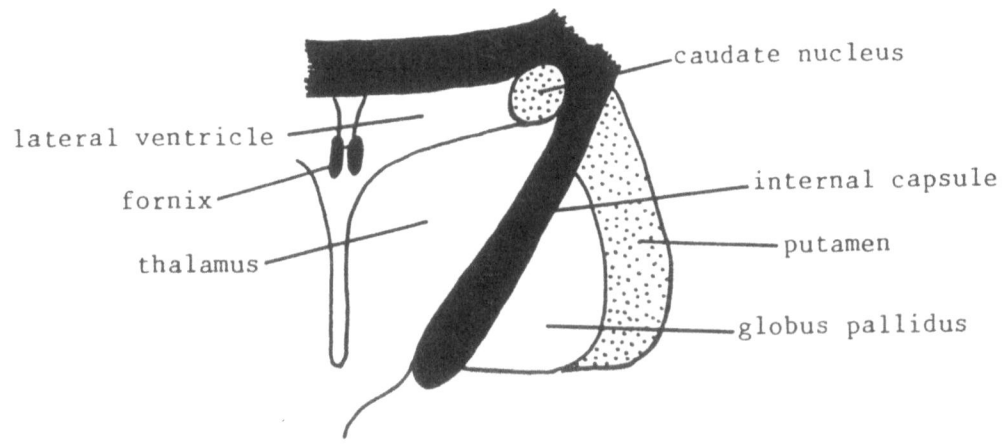

Coronal Section Through the Diencephalon and
Basal Ganglia

Archistriatum - Amygdaloid nucleus.

Paleostriatum - Globus pallidus.

Neostriatum - Caudate nucleus and putamen.

Fibers from widespread regions of the cerebral cortex synapse on cells in all parts of the basal ganglia, but particularly on cells of the caudate nucleus and the putamen. (Corticostriatal fibers).

Input on basal ganglia from thalamic nuclei:
(Thalamostriatal fibers)

Largely from the lateral ventral
                 anterior ventral   } . . . . Thalamic nuclei
                 centromedian
                 intralaminar

Most of these nuclei give rise to fibers which synapse upon cells of the putamen.

The substantia nigra in the midbrain tegmentum provides a considerable input to the putamen and, to a lesser extent, the globus pallidus (Nigrostriatal fibers).

NOTE: Nigrostriatal neurons release dopamine as their neurotransmitter substance. The rostral portion of the substantia nigra projects primarily to the caudate nucleus. The putamen receives fibers from the caudal portion of the substantia nigra.

The bulk of the efferent fibers from the basal ganglia arise from the globus pallidus. This nucleus influences numerous other regions via three main fiber pathways.

These are named:
           1. The lenticular fasciculus.

           2. The subthalamic fasciculus.

           3. The ansa lenticularis.

NOTE: In older terminology these pathways are referred to as divisions of the ansa lenticularis, thus the lenticular fasciculus is sometimes called the dorsal division of the ansa lenticularis. Similarly, the subthalamic fasciculus may be referred to as the intermediate division of the ansa lenticularis, and the ansa lenticularis may also be called the ventral division of the ansa lenticularis.

The Lenticular Fasciculus

These fibers pass transversely across the internal capsule dorsal to the subthalamic nucleus. Some of these fibers then leave this bundle to form the <u>thalamic</u>

<u>fasciculus</u> which courses dorsolaterally to terminate in the anterior ventral thalamic nucleus.

NOTE: Between the lenticular fasciculus and the thalamic
           fasciculus lies a small region of the gray matter - the zona
           inserta. Some thalamic fasciculus fibers terminate in this
           nucleus.

The remainder of the fibers of the lenticular fasciculus travel caudally to terminate in the hypothalamus and the midbrain (tegmental nuclei, substantia nigra, reticular nuclei and the red nucleus).

Subthalamic Fasciculus

These fibers pass directly medially from the globus pallidus to terminate directly on the subthalamic nucleus.

Ansa Lenticularis

These fibers course anteromedially, passing below the anterior limb of the internal capsule, then pass caudally to terminate in the red nucleus and pre-rubral area.

NOTE: Again in the older terminology, there is yet another set of
           terms sometimes used in regard to the fibers issuing forth
           from the globus pallidus, namely:
                Forel's Field H2 - Lenticular fasciculus
                Forel's Field H1 - Thalamic fasciculus
                Forel's Field H - Prerubral area

Tegmentobulbar and tegmentospinal fibers arise from tegmental cells of the midbrain. Rubrobulbar and rubrospinal fibers arise from the red nucleus. These terminate largely on cells adjacent to motor nuclei in the brainstem (bulbar fibers) or spinal cord (spinal fibers) - similar to the termination of the pyramidal tract fibers.

Most of these fibers cross to the opposite side of the brainstem upon leaving their nuclei of origin in the midbrain.

Fibers descending into the spinal cord travel in the lateral funiculus of the spinal cord, just ventral to the fibers of the lateral corticospinal tract.

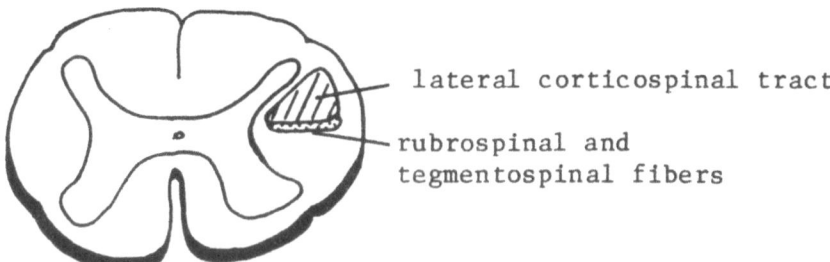

Descending Motor Fibers in the Spinal Cord

Central Tegmental tract fibers arising from tegmental cells (and some red nucleus cells) course caudally on the ipsilateral side of the brainstem, to terminate on the inferior olivary nucleus in the medulla. Olivocerebellar fibers arise from the inferior olivary nucleus and enter the cerebellum via the contralateral restiform body (inferior cerebellar peduncle).

Hence, basal ganglia information is relayed to the contralateral cerebellar cortex.

Movement Disorders

NOTE: Malfunctions of the basal ganglia system characteristically result in a group of diseases known generally as movement disorders or dyskinesias (athetoses, choreiform movements and resting tremors).

         These disorders may be expressed as too much movement such as occurs in hemiballism, for example, or alternatively, too little movement, as exemplified in the absence of facial expressive movements as seen in Parkinson's Disease.

         These dyskinesias are frequently accompanied by changes in muscle tone, which is usually increased.

Some Examples of Basal Ganglia Disorders

Hemiballism   Subthalamic nucleus lesions cause this disorder, which is characterized by wild flinging or thrashing movements of affected extremities on the opposite side of the body.

Parkinson's Disease   Caused by failure of the substantia nigra (and possibly other nuclei) to produce sufficient dopamine, or reduction in dopamine receptor sites in the lenticular nucleus. Mask-like facial expression, a stooping posture, resting tremor in the upper extremities ("pill rolling" tremor), gait disturbance, and a general reduction in associated movements characterize this disease. L-Dopa administration is frequently a successful means of treatment.

Huntington's Chorea   In this hereditary disorder, atrophy of the caudate nucleus is seen and eventually atrophy of cells in the cerebral cortex. Jerky choreiform movements of the extremities and axial musculature occur which are exaggerated under stress.

Wilson's Hepatolenticular Degeneration   Degeneration of the basal ganglia, particularly the lenticular nucleus. Accompanied by cirrhosis of the liver, and pigment deposit in the cornea near the junction with the sclera (Kayser-Fleischer Ring). Due to disturbances in copper metabolism (low serum ceruloplasmin) leading to excessive copper deposits in various tissues.

         Neurological signs are quite varied in this condition, but include bizarre movements, tremors and rigidity.

CEREBELLUM

The cerebellum is an important part of the motor control mechanisms within the central nervous system and exerts a considerable degree of influence upon this activity. It does not by itself initiate motor acts, but functions in a regulatory capacity, modifying the activities of other regions exerting control over motor function. The result of cerebellar modification of these areas is to ensure synergistic coordination of the various muscle groups involved in the performance of movements, thereby allowing these activities to be carried out smoothly.

In order to perform this function efficiently, the cerebellum receives input from many sources. These include sensory structures and pathways in addition to systems which influence lower motor neurons directly or indirectly (e.g. the pyramidal system and the basal ganglia system).

In the absence of normal cerebellar function, movements are still possible, although the manner of their performance may be markedly changed. The overall result of cerebellar malfunction may result in clumsiness and incoordination of movement.

The cerebellum is located in the posterior fossa lying posterior to the brainstem, and is covered superiorly by the tentorium. The cerebellum is comprised of cortex, deep white matter, and intrinsic (or deep) cerebellar nuclei. It is attached to the brainstem by three pairs of fiber bundles called the cerebellar peduncles. Grossly, the cerebellum consists of the narrow vermis on the midline, and two larger lateral masses, the cerebellar hemispheres which can be subdivided into lobes. These lobes are the posterior lobe (the largest), the anterior lobe and the flocculonodular lobe (the smallest).

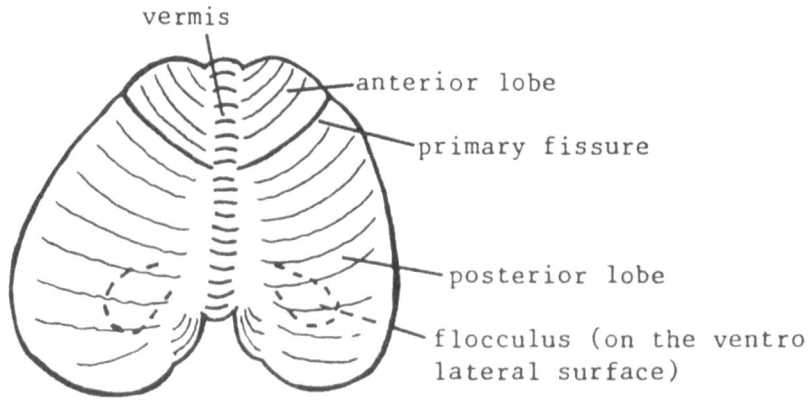

Cerebellum Viewed from Above

The flocculonodular lobe (archicerebellum) is associated with the vestibular system.

The anterior lobe (paleocerebellum) is largely involved in gross movements.

The posterior lobe (neocerebellum) helps in regulating the control of more discrete, fine motor movements.

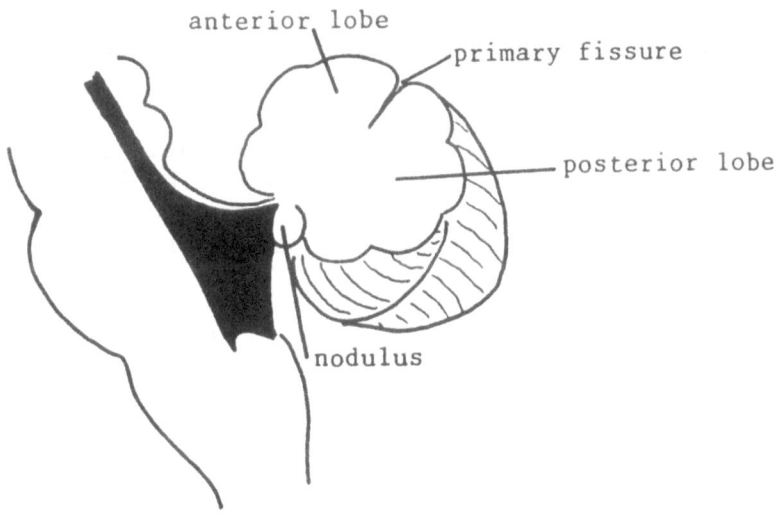

Midline Section through the Cerebellar Vermis

CEREBELLAR CORTEX

On the surface of the cerebellum the cerebellar cortex exhibits small convolutions which are called cerebellar folia.

Three histologic regions are found in the cortex:

1. The outer molecular layer (most superficial)
2. The inner granular layer (deeper and adjacent to the
white matter core)
3. The Purkinje cell layer (between the other two layers

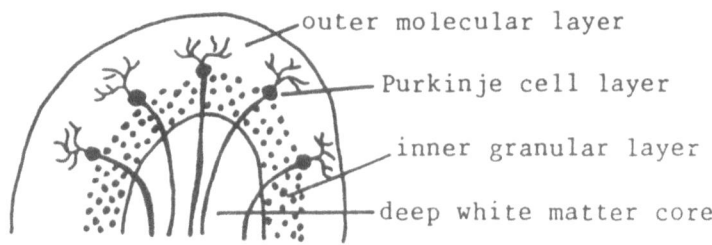

Cross Section of a Cerebellar Folium

Orientation of Cerebellar Cortex Cells in the Folium

Purkinje cells possess a profusely branched dendrite which extends into the molecular layer in a plane at right angles to the long axis of the folium. Axons of these cells leave the cell body and traverse the underlying granule layer, to become part of the white matter core. They may terminate on the deep cerebellar nuclei, or in other regions of cerebellar cortex.

Granule cells in the granular layer, possess numerous small dendrites surrounding their cell bodies. Their axons extend into the outer molecular layer, where they bifurcate and extend along the molecular layer, parallel to the long axis of the folium. This permits synaptic contact between these axons and the dendrites of numerous Purkinje cells.

The molecular layer contains the Purkinje dendrites and the branched axons of granule cells. Small neurons in this layer, called basket cells, have axons which run at right angles to the long folium axis, and synapse upon the bodies of Purkinje neurons.

Stellate cells also found in the molecular layer have axons which run longitudinally in the folium to synapse on the Purkinje cell dendrites.

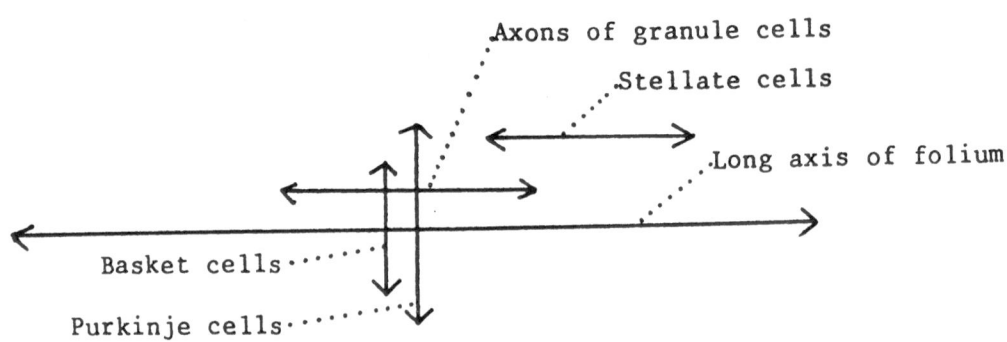

Orientation of Cells in the Cortex

Cerebellar Cortical Connections

Afferent fibers to the cerebellar cortex, arising from numerous sources, terminate as either:

    1. **Mossy fibers** - especially from pontine nuclei and the spinocerebellar systems, synapsing upon granule cell dendrites.

  or 2. **Climbing fibers** - especially from the inferior olive and reticular formation, synapsing upon Purkinje cell dendrites.

Purkinje cell axons are the only efferent fibers leaving the cerebellar cortex.

THE INTRINSIC OR DEEP CEREBELLAR NUCLEI

These consist of four paired nuclei embedded in the deep white matter, located superior to the roof of the fourth ventricle.

From medial to lateral they are named the:
> Fastigial nucleus (adjacent to the midline)
> Globose nucleus
> Emboliform nucleus
> Dentate nucleus (the largest cerebellar nucleus and most
> > laterally located).

These deep nuclei receive input from:
> 1. Purkinje Axons (all deep cerebellar nuclei).
> 2. Vestibular nuclei (fastigial nuclei).

The fastigial nuclei and a portion of the globose nuclei are functionally related to the flocculonodular lobe and the vestibular nuclei in the brainstem.

The remainder of the globose nucleus, the emboliform nucleus, and the large convoluted dentate nucleus, are the origin of the superior cerebellar peduncle (or brachium conjunctivum).

CEREBELLAR PEDUNCLES

Three pairs of fiber bundles provide the connection between the cerebellum and the brainstem and these are called the cerebellar peduncles.

1. Inferior cerebellar peduncle (also called the restiform body).
2. Middle cerebellar peduncle (the brachium pontis).
3. Superior cerebellar peduncle (the brachium conjunctivum).

All these peduncles provide routes for input into the cerebellum from motor regulating areas and also from numerous sensory systems. Cerebellar output for the most part is conducted via fibers of the brachium conjunctivum.

NOTE: Although sensory information from all sorts of receptors is
            relayed into the cerebellum, the cerebellum has no part in
            conscious awareness of sensory perception.

Components of Cerebellar Peduncles

1. Restiform Body

Afferent Fibers:
> I. Dorsal Spinocerebellar Tract – from the homolateral nucleus
> dorsalis in the spinal cord.

> II. Direct Arcuate Fibers – from the homolateral accessory
> cuneate nucleus in the upper medulla.

III. Olivocerebellar Fibers - from the contralateral inferior olivary nucleus.

IV. Trigeminocerebellar Fibers - from ipsilateral and contralateral nuclei interpolaris and rostralis (rostral portion of the descending spinal nucleus) of the trigeminal system.

V. Reticulocerebellar Fibers - from all parts of the medullary reticular formation.

VI. Arcuocerebellar Fibers - from nuclei in the medullary pyramids (actually these are displaced pontine nuclei).

Efferent Fibers:

Fastigiobulbar fibers arising from the fastigial nuclei, and making connections with the vestibular nuclei in brainstem. (These fibers travel in a compact bundle on the medial side of the restiform body known as the juxtarestiform body. The juxtarestiform body also carries afferent fibers from the vestibular system, comprised of some primary fibers from the vestibular ganglion and some secondary fibers from the inferior and medial vestibular nuclei).

2. Brachium Pontis

This is an entirely afferent peduncle, containing fibers arising from the contralateral pontine nuclei (part of the cortico-ponto-cerebellar circuit), located in the basal portion of the pons.

3. Brachium Conjunctivum

Afferent Fibers:
I. Ventral Spinocerebellar Tract - mostly from the ipsilateral nucleus dorsalis of the spinal cord, with a small contribution from the contralateral nucleus dorsalis.

II. Trigeminocerebellar Fibers - from the main sensory trigeminal nuclei (ipsilateral and contralateral).

III. Tectocerebellar Fibers - from the superior and inferior colliculi of the midbrain (ipsilateral and contralateral).

Efferent Fibers:

Arising from the dentate, emboliform and part of the globose nuclei. Fibers travel in the superior peduncle and decussate in the midbrain tegmentum. Most fibers after decussation course through and around the red nucleus in the midbrain tegmentum.

Some fibers synapse on cells of the red nucleus. However, most fibers continue forward to synapse upon cells of the lateral ventral nucleus of the thalamus, and also the centromedian thalamic nucleus.

NOTE: Some of the brachium conjunctivum efferent fibers will take
            a short caudal course after decussation to terminate in the
            reticular formation of the pons and medulla.
            Most, however, proceed rostrally after decussation.

CEREBELLAR FUNCTIONS

1. Coordination of somatic motor activity.
2. Regulation of muscle tone.
3. Influence on equilibrium mechanisms.

Vermis: Extensor muscle tone.

Paravermal Region of Hemispheres: Flexor muscle tone.

Lateral Region of Hemispheres: Influence upon skilled voluntary movement.

NOTE: Hemispheric cortical lesions may produce insignificant
            neurological signs. However, deeper lesions, especially if
            the dentate nucleus is affected, produce severe and lasting
            deficits.

Results of Cerebellar Lesions

1. If lateralized, produce ipsilateral signs (i.e. signs of cerebellar malfunction
   resulting from a right sided cerebellar lesion will be seen on the right side of
   the body).

2. Muscle tonus may be increased, with hyperactive reflexes. However, muscle tone
   is more commonly decreased with reduced, pendular reflexes.

3. Asynergy of affected muscle groups, expressed as:

   Dysmetria (mismeasurement of distance and failure of finger placement to
   accurately hit the mark, as in finger to nose testing, for example)

   Rebound Phenomenon (failure to adjust to rapid changes in muscle tension.
   Patient's arm is flexed and held against resistance; when resistance is
   stopped, the patient's arm flexes strongly due to inability to respond quickly
   enough to cessation of resistance to flexion).

   Dysdiadochokinesis (inability to perform accurately a series of rapidly
   performed, alternating movements, e.g., rapid supination and pronation of
   hands).

4. Intention Tremor (seen when attempting voluntary movement, not to be confused
   with resting tremors as seen in Parkinson's Disease).

5. Nystagmus (abnormal extraocular eye movements).

6. Ataxia of Speech (may be explosive, scanning or monotonous speech).

7. Gait Disturbance (broad based, "drunken" gait).

8. Trunkal Ataxia (especially with midline lesions, as for example, midline medulloblastoma of children).

Cerebellar cortical atrophy may occur due to a number of causes, and may give rise to any of the above symptoms.

SUMMARY OF MOTOR DEFICITS

Muscle strength is decreased with disorders of <u>muscle</u>, the <u>myoneural</u> junction and <u>lower</u> <u>motor</u> neurons. The degree ranges from mild weakness to paresis (more pronounced weakness) or paralysis, depending upon the extent and stage of the particular disorder.

Muscle stretch reflexes are <u>reduced</u> in all cases, as is muscle tone. Flaccidity is present (the ultimate in hypotonia) in cases of extensive lower motor neuron involvement.

In early stages of lower motor neuron disease <u>fasciculations</u> (spontaneous muscle twitches) may be readily observed.

NOTE: Occasional fasciculations can be observed in normal individuals which may be due to fatigue or compromise of blood supply and are <u>not</u> due to lower motor neuron disorder). Contractions of individual muscle fibers or <u>fibrillations</u>, cannot be observed by the eye, but can be detected with electromyography (E.M.G.). Atrophy of muscle is apparent in lower motor neuron disorders and generally present in muscle disease.

Nerve conduction studies and E.M.G. are of great diagnostic help in differentiating between disorders of lower motor neurons versus primary diseases of muscle.

With <u>upper</u> <u>motor</u> neuron disorder muscle strength also ranges from mild weakness to complete paralysis. Testing muscle stretch reflex response reveals <u>hyperactive</u> reflexes accompanied by increased muscle tone. The combination of hyperreflexia and hypertonia is sometimes referred to as <u>spasticity</u>. Muscle atrophy is generally not noticeable initially but may appear later due to disuse of the affected musculature. Although deep reflexes are exaggerated, superficial reflexes, such as the abdominal reflex, are diminished. In the lower extremity a <u>Babinski</u> reflex may be elicitable (upgoing great toe in response to plantar stimulation). Similarly <u>clonus</u> may be observed as a result of the hyperreflexia (sustained muscle contractions following muscle stretch).

NOTE: With spinal shock and sometimes following large cerebral stroke, there is a period when no reflexes are seen and flaccidity is present. Once this period has passed then the reflexes return and are hyperactive.

<u>Basal</u> <u>ganglia</u> system disorders do not generally affect motor strength or muscle stretch reflexes. "Lead pipe" rigidity may sometimes be seen (particularly in Parkinson's disease). Muscle atrophy is <u>not</u> a characteristic of these disorders.

Abnormal movements are characteristic of basal ganglia disease, too <u>much</u> movement being present (dyskinesia) as in <u>choreas</u>, <u>athetoses</u> and <u>ballism</u>, or too <u>little</u> (akinesia) as in the lack of facial expression for example, in Parkinson's disease. Muscle tremors at rest are also sometimes present ("pill rolling" in Parkinson's).

Cerebellar disorders give rise to <u>intention</u> <u>tremor</u> (not at rest, but evident on attempting voluntary movement). Imbalance between opposing muscle groups is the

underlying cause of intention tremor due to the incoordination and assynergy present
with cerebellar lesions. This also leads to dysmetria (overshooting the mark) and
ataxia (clumsiness). Difficulty in performing rapidly alternating movements
(adiodochokinesia) and rebound (failure to stop flexion or extension against force
once the force is removed) also result from this muscle imbalance and
incoordination.

Hypotonia is sometimes present and muscle stretch reflexes are sometimes pendular.

THE VISUAL SYSTEM

Receptors for vision are found in the retinal layer of the eye. Light entering the eye passes through the transparent cornea and the pupillary opening in the iris diaphragm, then through the lens to be focussed on the retina. Clear focus of light rays is achieved by refraction occurring in the cornea and the lens. The amount of light admitted to the retina is regulated by the size of the pupillary aperture, which is governed by the autonomic nervous system. Pupillary constrictor muscles in the iris are controlled by parasympathetic fibers of the oculomotor nerve, whereas pupillary dilator muscles are activated by the sympathetic nervous system.

Movement of the eye within the orbit is accomplished by the activity of extraocular eye muscles which insert onto the sclera, the outer layer of the globe. Simultaneous adjustments of the position of both eyes to result in normal vision requires a considerable degree of central nervous system regulation. The ability to achieve normal eye movements may be impaired subsequent to malfunction of various portions of the C.N.S., ranging from cerebral cortex to brainstem or the three cranial nerves (oculomotor, trochlear, abducens) controlling the extraocular muscles.

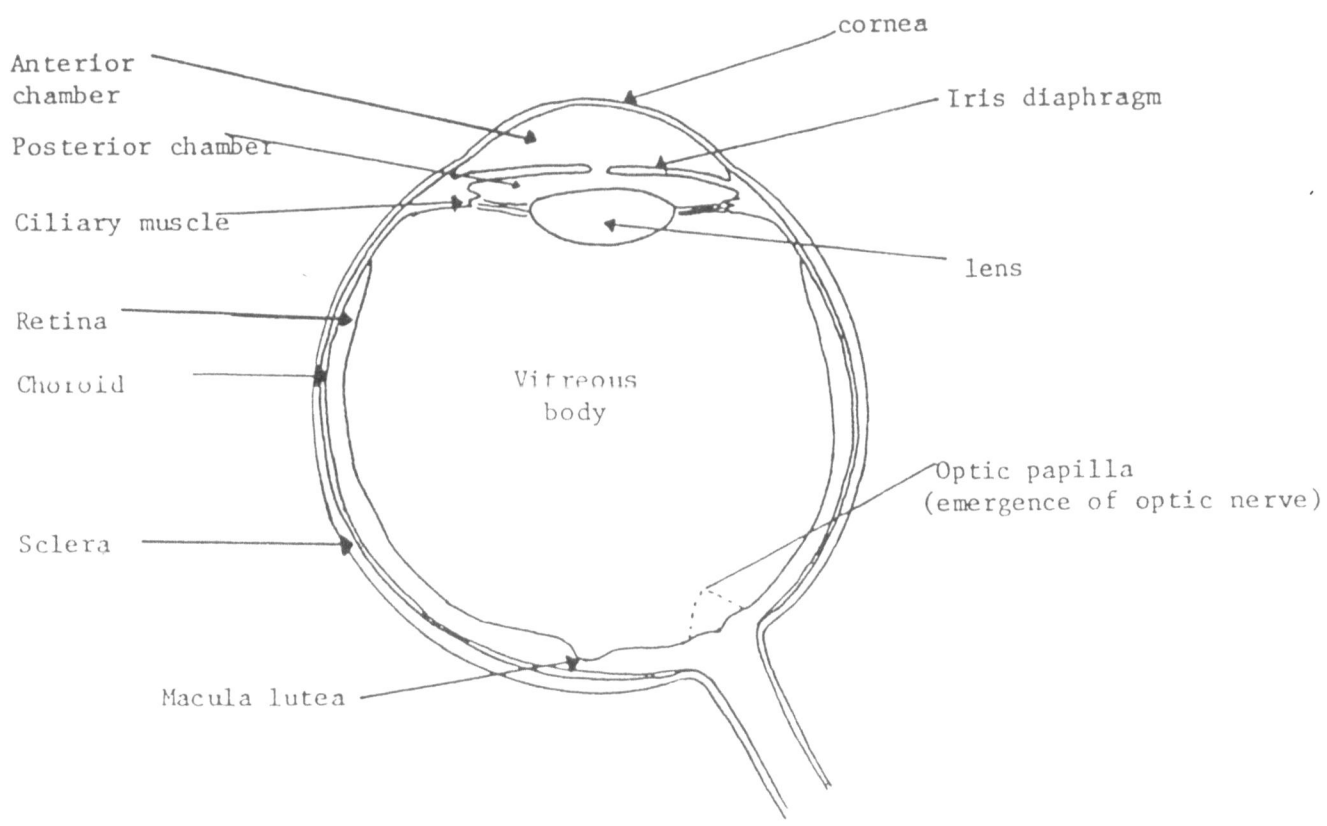

Horizontal Section through the Eye

Retina

The neuronal elements of the retina are shown in the diagram. Modified bipolar neurons, the rods and cones, constitute the light sensitive components. Their distribution throughout the retina varies with the region examined. The <u>fovea centralis</u> contains cones only, which respond to daytime high intensity illumination, for sharp vision and color discrimination. Cones and some rods are seen in the surrounding region, the <u>macula lutea</u>.

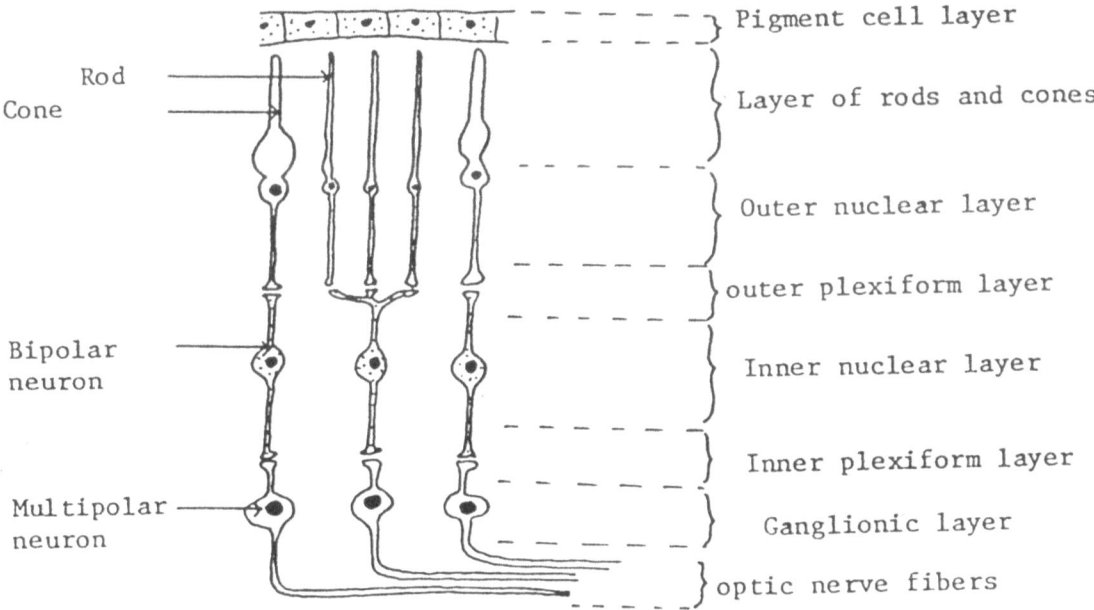

Arrangement of Nerve Cells in the Retina

The remainder of the retina contains only rods responding to peripheral vision, low intensity illumination, and are particularly important for night vision.

Light strikes the retina, and the resultant photochemical response initiated at the pigment cell layer activates rods or cones. Impulses are relayed via bipolar neurons to ganglion cells, axons of which constitute the optic nerve fibers, which are not myelinated whilst still within the retina.

NOTE: Light entering the eye passes through all the layers of the retina before it reaches the pigmented cell layer. The arrangement of the retina is such that light falling on the macula and fovea passes through the least amount of cellular elements before striking the light sensitive portion, and this region is responsible for the best visual acuity.

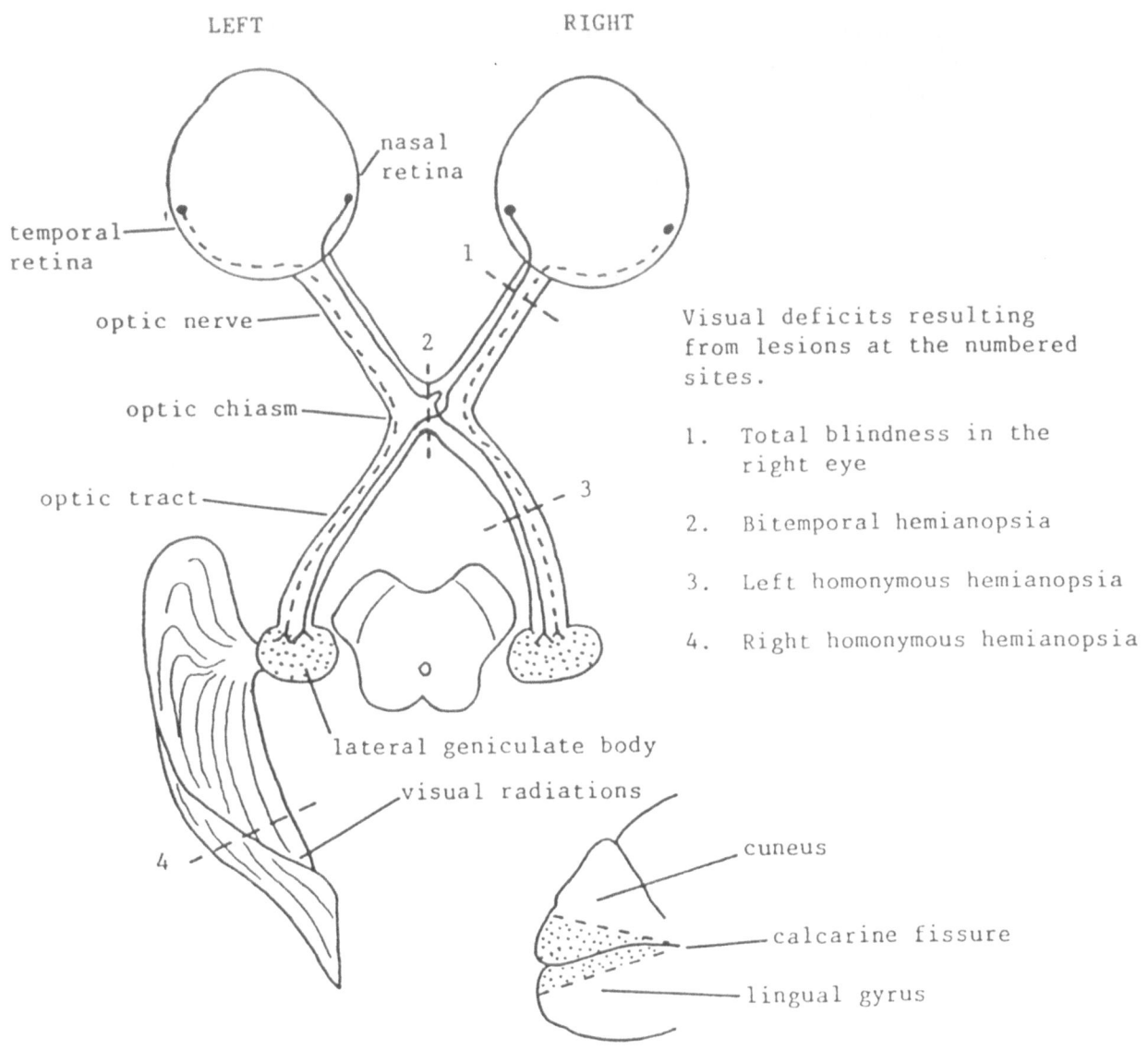

LEFT   RIGHT

nasal
retina

temporal
retina

optic nerve

optic chiasm

optic tract

lateral geniculate body

visual radiations

cuneus

calcarine fissure

lingual gyrus

Visual deficits resulting
from lesions at the numbered
sites.

1. Total blindness in the
    right eye

2. Bitemporal hemianopsia

3. Left homonymous hemianopsia

4. Right homonymous hemianopsia

The Visual Pathways

Optic Nerve

Fibers of the optic nerve become myelinated once the nerve leaves the globe, but
because it is really an extension of the diencephalon, the myelin is formed by
oligodendrocytes rather than Schwann cells.

At the immediate point of exit from behind the eyeball, macular fibers lie on
lateral side of the optic nerve, but rapidly assume a position in the center of the
nerve surrounded by paramacular fibers and peripheral retinal fibers on the outside.

NOTE: The entering central retinal artery and the exiting vein
            occupy a position in the center of the optic nerve, at the
            point of its attachment to the retina (the optic papilla).

During ophthalmoscopic examination of the retina, the optic papilla (optic disc) should be observed closely, as a variety of changes in appearance may accompany pathologic conditions either intracranially or elsewhere. For example, elevated intracranial pressure (especially of a chronic nature) may result in a swollen papilla and changes appearance of the vessels.

Optic Chiasm and Optic Tract

The optic nerves travel intracranially via the optic foramina towards their attachment to the diencephalon. Prior to this attachment there is a partial decussation (the optic chiasm) occurring superior to the diaphragma sella and just anterior to the pituitary stalk. At the chiasm, fibers from the nasal half of the retina cross the midline to join with the uncrossed temporal fibers of the opposite eye to form the optic tract. Thus, posterior to the chiasm, an optic tract is comprised of fibers arising from the temporal half of the retina of the ipsilateral eye and also fibers from the nasal half of the retina of the contralateral eye. At the chiasm, fibers crossing in the anterior portion are from the inferior part of the nasal retina. Fibers from the superior portion of the nasal retina cross in the posterior part of the chiasm.

NOTE: Some nasal fibers may loop into the contralateral optic nerve or into the ipsilateral optic tract for a few millimeters before completing their crossing into the contralateral optic tract. These fibers are referred to as Von Willebrand's Knee.

The optic tracts pass around the lateral surface of the cerebral peduncles towards their diencephalic destination which is the lateral geniculate body of the thalamus.

Lateral geniculate Body

Rotation of optic tract fibers occurs prior to termination in the lateral geniculate body. Fibers arising from the superior quadrants of both eyes (superior temporal of the ipsilateral eye, and superior nasal of the contralateral eye) synapse in the medial part of the lateral geniculate body. Conversely, inferior retinal quadrant fibers terminate in the lateral portion.

Macular fibers are thought to terminate in a cone with the apex of the cone towards the central portion of the lateral geniculate body and the base of cone towards the superior portion.

NOTE: Some fibers of the optic tract (probably collateral branches of optic tract axons) continue beyond the lateral geniculate body without synapse there, as the brachium of the superior colliculus. These terminate either on neurons of the superior colliculus or alternatively on neurons of the pretectal region. Pretectal region connections mediate pupillary light reflexes.

Superior colliculus connections mediate reflex movements of

the eyes and head in response to visual stimuli.

Visual Radiations (Geniculocalcarine Tract)

The visual cortex (calcarine cortex) is comprised of the cuneus and lingual cortex
on the medial side of the occipital lobes. The visual radiations are the fibers
passing from the lateral geniculate body to the visual cortex.

Fibers arising from the medial portion of the lateral geniculate body, representing
the superior retinal quadrants, terminate on the superior lip of the calcarine
fissure, i.e. the cuneus. The lateral portion of the lateral geniculate body
(inferior retinal quadrants) projects to the lower lip of the calcarine fissure
(lingual cortex).

The macular retinal representation is in the posterior portion of the visual cortex.

Visual radiations from the medial portion of the lateral geniculate body (superior
retinal quadrants) pass almost directly posteriorly to terminate in the cuneus.
Radiations from the lateral part of the lateral geniculate body (lower retinal
quadrants) pass forward a considerable distance into the temporal lobe before
looping posteriorly over the temporal horn of the lateral ventricle to terminate in
the lingual cortex. These temporal lobe radiations are sometimes called the Loop of
Meyer.

NOTE: A Pearl!   The Five "L's" –
                  Lower retinal fields
                  Lateral portion of L.G.B.
                  Lowermost fibers of visual radiations
                  Loop of Meyer
                  Lingual cortex

REMEMBER! The retinal fields are the opposite of visual fields. For example, lower
(inferior) retinal fields are responsive to upper (superior) visual fields. In
other words, light originating from above the visual horizon (superior visual field)
falls on the inferior portion of the retina.

Unilateral lesions of the visual pathway posterior to the chiasm will result in
contralateral visual field deficits. For example, a lesion of the left side optic
tract produces a right side visual deficit.

NOTE: The general term employed for a visual field deficit is
            anopsia. Thus a left side anopsia means the individual
            cannot perceive light arising from his or her left when
            looking straight ahead. If the visual deficit applies to
            both eyes (i.e. light from the left visual field falling
            upon nonresponsive nasal retina of the left eye and
            nonresponsive temporal retina of the right eye) then the
            deficit may be called a left homonymous hemianopsia.
            Partial unilateral lesions of the visual radiations or
            lesions affecting a portion of the visual cortex in one
            hemisphere, may produce quadrantic visual field deficits
            (quadranopsia). Thus interruption of the lowermost visual

radiations in the left hemisphere (Loop of Mayer) cause a <u>right</u> <u>sided</u> <u>superior</u> <u>quadranopsia</u>. In this situation individuals would not perceive light originating from their upper right side when looking straight ahead. (i.e. a "pie in the sky" visual defect).

OCULOMOTOR SYSTEM

Movements of the eyes within the orbit are brought about by the action of the six extraocular eye muscles which are inserted into the globe. These muscles, their innervation and action, are listed below.

MUSCLE	NERVE	ACTION
Superior Oblique	C.N. IV	Abduction, depression intorsion
Inferior Oblique	C.N. III	Abduction, elevation, extorsion
Superior Rectus	C.N. III	Elevation, adduction intorsion
Inferior Rectus	C.N. III	Depression, adduction extorsion
Medial Rectus	C.N. III	Adduction
Lateral Rectus	C.N. VI	Abduction

In normal circumstances, transfer of gaze whilst maintaining constant head position requires movements of both eyes simultaneously in order to perceive a single visual image. These paired movements of the eyes within the orbits can be called conjugate eye movements. It must also be recognized that eye position is the result of the combined action of all extraocular muscles working together. This requires an exquisite degree of balance and control of the six muscles regulating the position of each eye. Malfunction of a variety of regions within the C.N.S. may result in interference with this delicately regulated mechanism.

In the presence of cranial nerve palsies affecting the regulation of some of the extraocular eye muscles, gaze transfer may be compromised such that the eyes do not both move in a conjugate manner. Thus both eyes do not view the object of gaze in precisely the same way, which can result in a double image (diplopia).

OCULOMOTOR NERVE (C.N. III)

Nucleus - located in the midbrain periaqueductal grey matter ventral to the aqueduct, dorsal to the medial longitudinal fasciculus, at the level of the superior colliculus. The oculomotor nuclear complex consists of a number of individual subnuclear groups.

1. Paired nuclei.

   A. The lateral nucleus (somatic cell group) is the major nucleus consisting of 4 divisions which innervate the superior, inferior, medial rectus muscles and the inferior oblique muscle, respectively.

NOTE: Superior rectus muscle innervation is both crossed and uncrossed, i.e., some fibers cross over to become part of

the contralateral nerve to the superior rectus of the
contralateral eye.

B. The Edinger-Westphal nucleus (visceral cell group) is located rostral to the
lateral group. This is the origin of parasympathetic preganglionic fibers
which cause pupillary constriction and also regulation of lens curvature via
the ciliary muscles.

2. Midline nuclei

A. A caudal central nucleus which provides fibers innervating the levator
palpebrae superioris muscle in the upper eyelid.

B. Nucleus of perlia – "Convergence" nucleus.

3. Association nuclei

A. Nucleus of Darkschewitsch – associated with vertical eye movements. Fibers
from this nucleus project into the posterior commissure.

B. Interstitial nucleus of Cajal – associated with vertical eye movements.
Fibers from this nucleus project via the posterior commissure to synapse on
the somatic cell groups of the oculomotor group and also onto the nucleus of
the trochlear nerve.

The Rostrocaudal Arrangement of the
Nuclei in the Oculomotor Complex

Peripheral Course

A. Subarachnoid The oculomotor nerve emerges from the interpeduncular fossa,
passing between the posterior cerebral and superior cerebellar arteries.

B. Cavernous sinus portion – enters lateral to the posterior clinoid process,
and is the most superior nerve in the lateral wall of the sinus.

C. Passes through the superior orbital fissure to enter the orbit.

D. In the orbit the nerve gives muscular branches to the superior rectus and the superior palpebral muscles, the medial and inferior rectus, and the inferior oblique muscles. The parasympathetic portion innervates the ciliary and pupillary constrictor muscles (after synapse in the ciliary ganglion).

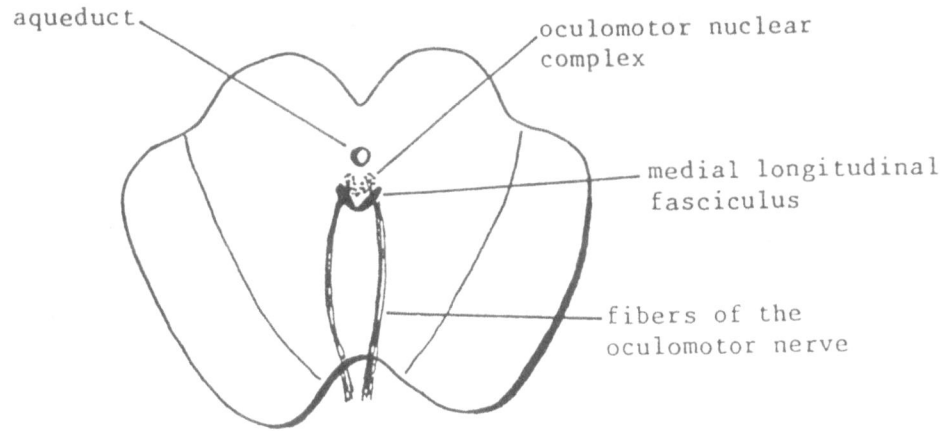

Cross Section through Midbrain at the Level of the Superior Colliculus

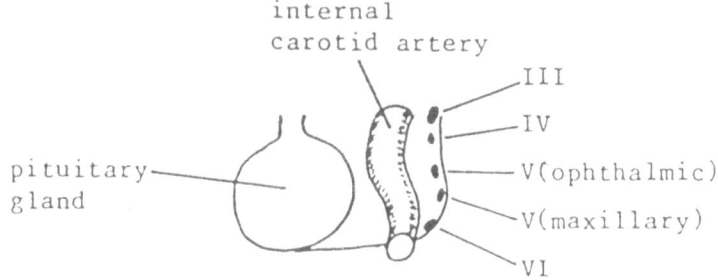

Section through the Cavernous Sinus showing the Position of the Nerves in the Wall

PUPILLARY LIGHT REFLEXES

The size of the pupillary aperture is governed in a reflex fashion by activity of the autonomic nervous system and is influenced, in normal circumstances by the amount of light reaching the retina. Thus high intensity light (e.g. being in bright sunlight) results in reduction of pupil size for protection of the retina against overstimulation. Conversely moving into a darkened room from a region of high intensity illumination, brings about reflex dilatation of the pupil.

A bright light shone into one eye should result in reflex reduction in pupil size of that eye. This is called the _direct_ pupillary reflex.

At the same time, this stimulus should also result in the reflex reduction in pupil size of the contralateral eye. This would be the _indirect_ pupillary reflex in regard to that eye.

This reflex uses the optic nerve for the afferent limb, and the parasympathetic portion of the oculomotor nerve for the efferent limb. The response in both pupils after the stimulus is applied to only one eye may also be termed the consensual pupillary response.

NOTE: Lesions of the posterior commissure are said to reduce or even eliminate the consensual light reflex.

Lesions of the Third Cranial Nerve may cause:

- Ipsilateral external strabismus (abduction of the eye) due to the unopposed action of the intact lateral rectus muscle.

- Ptosis (lowering of upper eyelid due to weakness of the superior palpebral muscle).

- Absence of upward or inward (toward the nose) movements of the eye.

- Lesions causing compression of the superior colliculus (for example, tumors in the region of the pineal organ) may cause paralysis of vertical gaze.

- Dilation of pupil (mydriasis) because of loss of parasympathetic innervation.

- Absence of the pupil light reflex, also due to the loss of parasympthetics.

- Absence of convergence and accommodation which would normally occur when gazing at close objects.

- Lateral diplopia on gazing to the side opposite the lesion.

TROCHLEAR NERVE (C.N. IV)

Nucleus – In the midbrain periaqueductal grey matter, ventral to the aqueduct, dorsal to the medial longitudinal fasciculus, at the level of the inferior colliculus.

NOTE: Axons leaving the nucleus decussate dorsal to the aqueduct before exiting the brainstem dorsally, just caudal to the inferior colliculus. Thus the fibers of the left trochlear nerve arise from the trochlear nucleus on the right side of the midline.

Peripheral Course

The trochlear nerve passes around the midbrain between the posterior cerebral and superior cerebellar arteries, close to the inferior surface of the tentorium.

It then enters the lateral wall of the cavernous sinus (see diagram), and hence through the superior orbital fissure into the orbit to supply the superior oblique muscle.

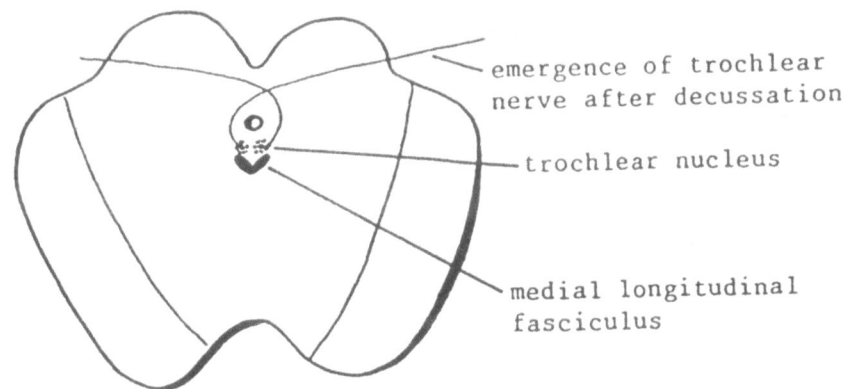

**Midbrain Section at the Level of the
Inferior Colliculus**

Lesions

It is difficult to detect isolated fourth nerve lesions. However, they may cause
vertical diplopia, especially on downward gaze to the side opposite the lesion. The
patient may tilt his or her head towards the side opposite to the lesion

ABDUCENS NERVE (C.N. VI)

<u>Nucleus</u> - Just below the floor of the 4th ventricle at the junction of the pons and
the medulla

Peripheral Course

Exits ventrally at the pontomedullary sulcus. Passes forward across the base of the
pons to penetrate the cavernous sinus lateral to the internal carotid artery (see
diagram) to the superior orbital fissure, into the orbit, then to the lateral rectus
muscle.

NOTE: With unilateral lesions of abducens nerve fibers, the
              affected eye is adducted (turned towards nose) due to
              paralysis of the lateral rectus muscle. The other eye is
              not affected.

Unilateral lesions of the abducens nucleus and adjacent para-abducens nucleus,
results in both eyes being deviated to the side opposite the lesion (lateral gaze
paralysis syndrome).

REASON: Loss of innervation of lateral rectus on the affected side,
              plus absence of coordination from the para-abducens nucleus
              to the contralateral oculomotor complex via the medial

longitudinal fasciculus (especially the nuclei regulating the medial rectus muscle), resulting in contralateral medial rectus palsy. Hence, both eyes are deviated to the side opposite the side of the lesion.

CONTROL OF EYE MOVEMENTS via NUCLEI OF CRANIAL NERVES III, IV, and VI

Vestibular Regulation

There is a considerable amount of vestibular input into the medial longitudinal fasciculus for reflex eye movements in response to positional changes of the head and neck. Vestibular stimulation giving rise to nystagmus movements. (Caloric tests, etc.)

Doll's eye Reflex - With the patient supine (lying on his back), if the head is held so the eyes look straight ahead, and then rotated to the right, the eyes move towards the patient's left, so as to continue looking towards the observer. If the head is moved to the left the eyes move to the right. Similarly, rotating the head up causes the eyes to look down and vice versa. Indicates the brainstem mechanisms for regulating eye movements are intact, but suggests cerebral malfunction. If this movement results in disconjugate, random movements of the eyes, it is suggestive of more severe damage affecting the brainstem.

Cortical Eye Fields

Two regions - (i) frontal (posterior portion of the middle frontal gyrus),
(ii) occipital (lateral occipital cortex).

Frontal Eye field

Stimulation produces lateral conjugate eye movements toward the side contralateral to the stimulation side.

Destructive lesions cause deviation towards the side of lesion. May relay to eye muscle nuclei via the basal ganglia.

NOTE: Stimulation of frontal eye fields may also give rise to vertical eye movements, either up or down.

Occipital Eye Field

Stimulation causes conjugate movements movements to the opposite side. Lesions cause deviation to the same side as the lesion. Pathways through internal capsule to the midbrain region.

Subcortical areas influencing eye movements:

Basal Ganglia   - stimulation in some areas produces deviation of the eyes to the opposite side.

Pontine Regions   - dorsal pontine tegmentum, adjacent to the abducens nucleus (para-abducens nucleus).

                                Stimulation causes conjugate deviations of the eyes towards the same side as the side stimulated.

Superior Colliculus - stimulation may give rise to vertical eye movements

VESTIBULAR AND COCHLEAR SYSTEMS

The eighth cranial nerve could, in a sense, be regarded as two separate nerves responsible for two different special senses. The vestibular portion of the nerve transmits information into the C.N.S. arising from receptors responsive to movement and gravitational force. The sense of hearing is of course the role of the cochlear nerve. Although the two parts of the eighth nerve have obviously different functions, there are several characteristics which both portions have in common. They are both special senses, their receptor structures are similar in some regard and the two divisions of the nerve are closely related to each other in their course from receptors to the brainstem.

THE VESTIBULAR SYSTEM

This elaborate sensory system is concerned primarily with mediating reflex adjustment of position of the eyes, head, neck, trunk, and extremities in response to movement and changes in head position relative to gravitational force. It is thus of importance in helping to maintain balance, equilibrium and orientation relative to the surroundings.

Receptors

Located in the inner ear, they consist of two types: (1) Crista Ampullaris, found in the three semicircular canals (dynamic receptors); (2) Maculae of the utricle and saccule (static receptors).

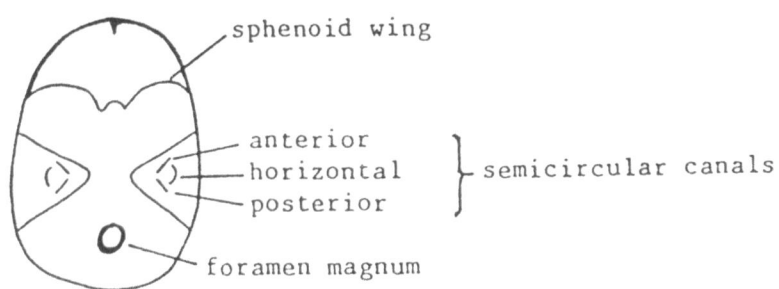

Base of the Cranium showing the Position of the
Semicircular Canals in the Petrous Bone

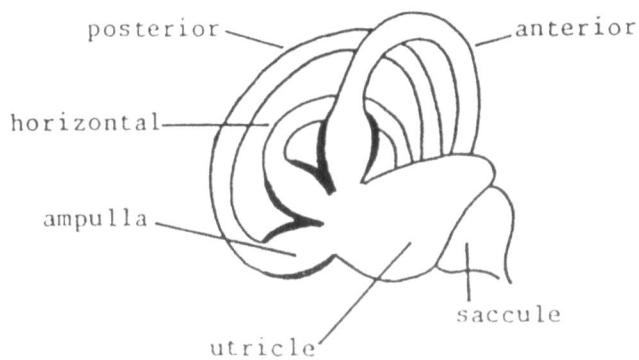

The Arrangement of the Vestibular Labyrinth

In the interior of the petrous portion of the temporal bone resides an elaborate arrangement of membrane lined canals disposed in three planes. The canals contain otic fluid (endolymph) which is displaced with rotational movements of the head (angular acceleration). Maximal displacement occurs in whichever canal is at right angles to the axis of rotation.

In the dilated ampulla portion of the canals are found the receptors of this system, the cristae ampullaris. The cristae, one for each canal, consist of a gelatinous cupula with hair cells embedded within, which can be distorted by the movement of otic fluid. This distortion initiates neural impulses in the vestibular fibers innervating these structures.

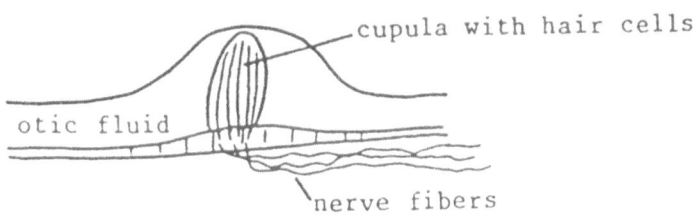

Crista Ampullaris in the Semicircular Canal

The maculae, static receptors (responding to gravitational force and linear acceleration) in the utricle and saccule, consist of hair cells which are in contact with an overlying thin gelatinous membrane. Small calcium carbonate granules (otoliths) upon this membrane change position in response to gravity, hence distorting hair cells in a particular region of the macula. This in turn initiates activity in the nerves innervating the hair cells.

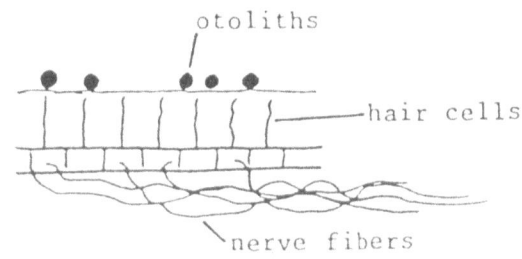

Utricula Macula

Peripheral processes of bipolar neurons located in the vestibular ganglion (Scarpa's Ganglion) found in the internal auditory canal, innervate the receptors. Central processes form the vestibular portions of the eighth nerve and traverse the internal auditory canal to attach to the brainstem at the cerebellopontine angle, ventral to the restiform body. After entering the brainstem, these fibers form ascending and descending roots, which terminate, for the most part, upon cells of the vestibular nuclei.

Four vestibular nuclei are located below the floor of the fourth ventricle.

The <u>lateral</u> vestibular nucleus is located at the pontomedullary junction, at the level of entry of the vestibular nerve. At the same level is found the <u>medial</u> vestibular nucleus.

The superior cerebellar puduncle forms the dorsolateral boundary of the <u>superior</u> vestibular nucleus, which is located at the level of the main sensory nucleus of the trigeminal nerve. The <u>descending</u> (or spinal) vestibular nucleus extends caudally to approximately the level of the accessory cuneate nucleus.

Some fibers of the ascending root enter the cerebellum via the juxtarestiform body, terminating upon the fastigial nucleus, and cortex of the flocculonodular lobe (direct vestibulocerebellar fibers).

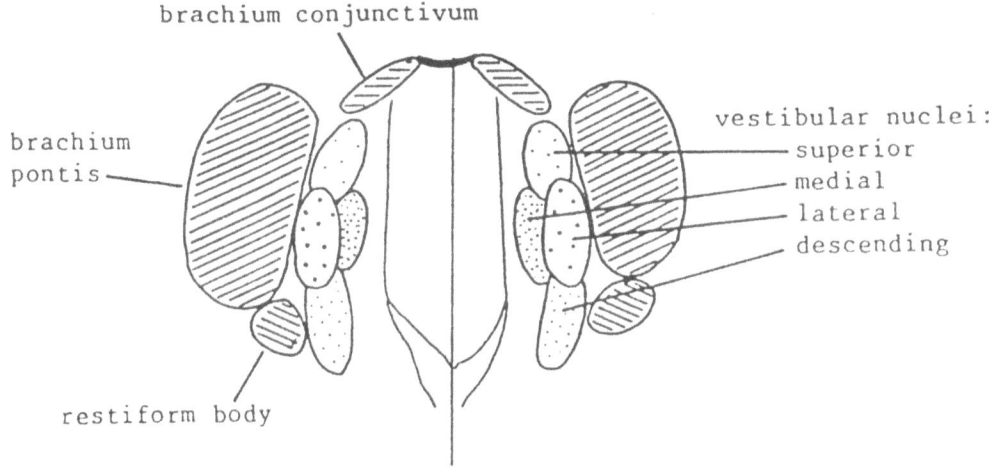

Floor of the Fourth Ventricle Indicating the
Relative Positions of the Vestibular Nuclei

Cerebellar efferent fibers arising from the fastigial nuclei also travel in the juxtarestiform body, synapsing on cells of all of the brainstem vestibular nuclei, and thereby subjecting them to some cerebellar modulation.

Central Connections of Vestibular Nuclei

1. Into the medial longitudinal fasciculus (M.L.F.).
2. Cerebellum.
3. Vestibulospinal tracts.
4. Reticular formation.

The medial longitudinal fasciculus extends from the rostral midbrain to the lowermost cervical spinal cord. M.L.F. fibers in the brainstem are located adjacent to the midline, just ventral to the ventricular system. In the spinal cord these fibers are found in the ventral funiculus. Vestibular input from all vestibular nuclei enters the M.L.F. bilaterally. Ascending M.L.F. fibers allow vestibular influence to be brought to bear on the abducens, trochlear, and oculomotor nuclei, thus influencing extraocular eye movements.

Descending fibers in the M.L.F. pass into the anterior funiculus of the spinal cord, terminating upon the intermediate gray in the cervical region. Thus providing vestibular modification of neck muscles and muscles in the upper extremity. These descending fibers in the M.L.F. are also called the medial vestibulospinal tract.

Some axons from vestibular nuclei neurons enter the cerebellum via the juxta-restiform body, with similar connections to the direct vestibulocerebellar fibers mentioned above.

The main descending vestibular pathway is the lateral vestibulospinal tract arising primarily from the lateral vestibular nucleus. This descends into all spinal cord levels coursing in the region of the junction between the lateral and anterior funiculi of the cord. This tract allows vestibular influence to reach musculature of the trunk and the lower extremity.

Diffuse projections into the reticular formation effect connections with visceral areas (visceral and autonomic reflexes in response to vestibular stimulation).

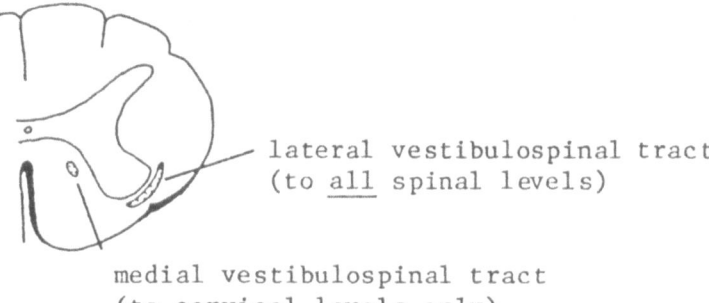

lateral vestibulospinal tract
(to all spinal levels)

medial vestibulospinal tract
(to cervical levels only)

The Position of the Vestibular Fibers
in the Spinal Cord

Vestibular projection to the cerebral cortex is thought to be via the medial lemniscus or by multisynaptic pathways in the reticular formation.

Lesions of the vestibular system may produce nystagmus, impaired balance, vertigo, nausea, vomiting, and sweating.

NOTE: Nystagmus - movements of the eyes with a rapid movement in .
          one direction and a slower recovery movement in the opposite
          direction. The direction of the nystagmus movement is named
          for the direction of the rapid phase of eye movement.

NOTE: Caloric stimulation, using either warm or cold water placed
          in the external auditory canal, may be employed to test the
          integrity of the vestibular and oculomotor systems. Cold
          water calorics should result in nystagmus eye movements with
          the rapid phase away from the side being stimulated (i.e.,
          cold water placed in the left ear should result in right
          beating nystagmus). Warm water calorics should result in

nystagmus with the rapid phase to the same side as that being stimulated.

In comatose patients with an intact brainstem, cold calorics produce only a slow drift of the eyes towards the side being stimulated. No response occurs in the presence of brainstem malfunction.

THE AUDITORY SYSTEM

The receptors of this system are found within a coiled canal (the cochlea) within the petrous bone. The cochlear spiral consists of two and a half turns around the bony modiolus. Membranes divide the cochlea along its extent into three compartments. These are the scala vestibuli and the scala tympani which communicate with each other at the apex, and the scala media located between the two (see diagram).

Auditory stimulation initiates vibration of the tympanic membrane which is transmitted via the ear ossicles (maleus, incus and stapes) to the oval window at the base of the cochlea. Vibration of the oval window causes pressure waves in the periotic fluid of the scala vestibuli which are transmitted to the fluid in the scala tympani. The basilar membrane then vibrates causing hair cells to come into contact with the tectorial membrane, initiating neural impulses.

The organ of Corti within the scala media of the cochlea of the inner ear, contains hair cells, which constitute the receptor structures.

Low notes cause vibration of the upper portion of the basilar membrane and are detected by the organ of Corti at that location.

High notes result in vibration in the lower turns and are detected by the lowermost portions of the organ of Corti.

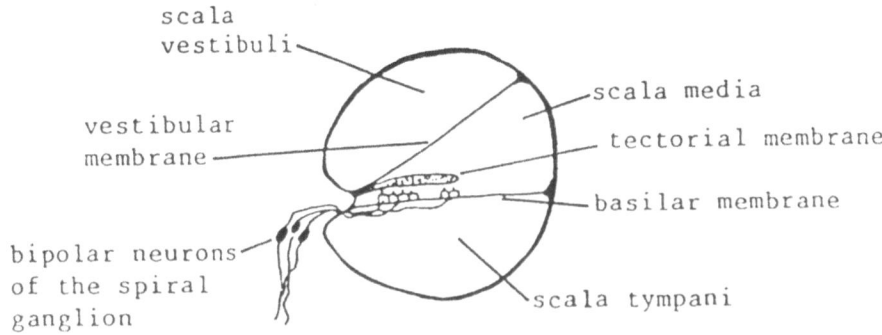

Section of the Cochlea

Bipolar neurons located in the modiolus constitute the spiral (or cochlear) ganglion. Peripheral processes innervate hair cells of the organ of Corti. Central processes constitute the cochlear division of the eighth nerve and traverse the internal auditory canal accompanied by the vestibular division, and attach to the brainstem at the cerebellopontine angle. Relationships of the auditory nerve within the internal auditory canal are as follows: Facial nerve and superior division of the vestibular nerve are above; the cochlear nerve and the inferior division of the vestibular nerve are below (separated by the transverse crest of the internal auditory canal).

Auditory fibers pass dorsal to the restiform body and terminate in the dorsal and ventral cochlear nuclei. (Cochlear nuclei lie on the dorsolateral aspect of the restiform body, and are continuous with each other).

Central Connections of the Cochlear Nuclei

Reflex connections between cochlear nuclei and the facial nerve nucleus mediate the stapedial reflex. Similar connections to the motor trigeminal nucleus mediate the tensor tympani reflex. These are protective reflexes which help to reduce the effects of excessive auditory stimulation.

The auditory relay pathway to the cerebral cortex involves the following groups of nuclei: Superior olivary nucleus; nucleus of the trapezoid body; nucleus of the lateral lemniscus; nucleus of the inferior colliculus; the medial geniculate nucleus.

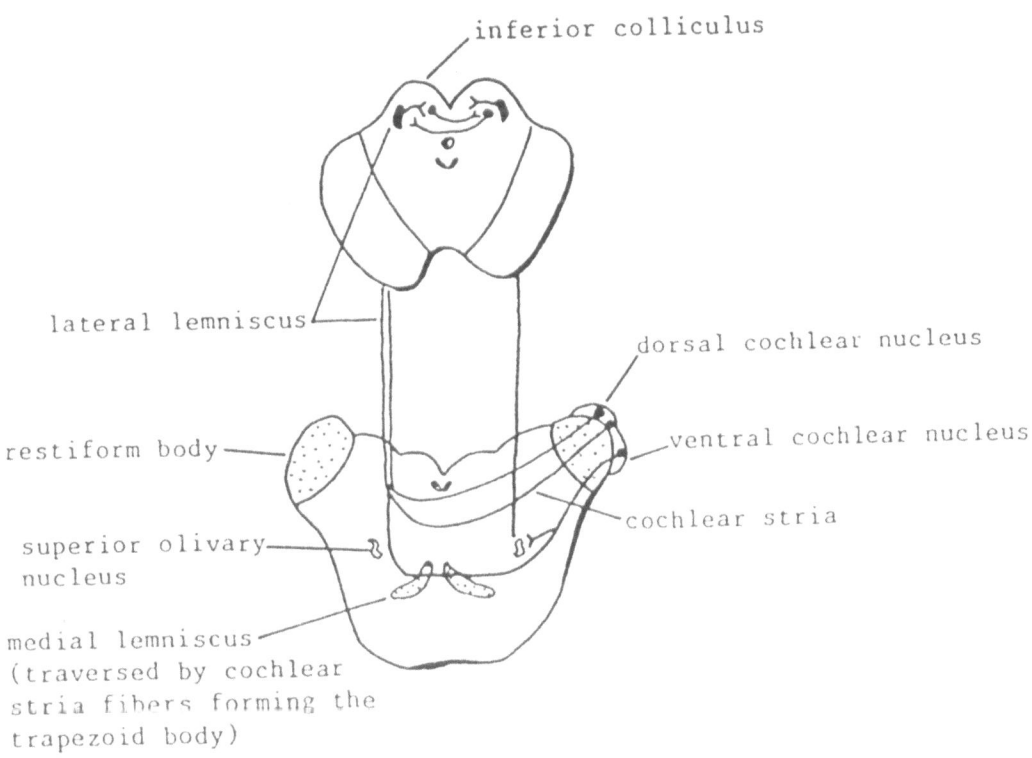

The Auditory Pathways

Axons leave both cochlear nuclei and cross the brainstem to form the lateral lemniscus on the contralateral side. These axons, in crossing the midline, form the cochlear striae (dorsal, intermediate and ventral). The ventral stria intermingle with ascending fibers of the medial lemniscus in their course across the midline, thereby forming the Trapezoid Body. Some cochlear fibers synapse upon neurons in this location (nucleus of the trapezoid body). Trapezoid body neurons project axons into the lateral lemniscus.

Intermediate stria axons may synapse on cells of the ipsilateral superior olivary nucleus. Cells of this nucleus project axons into the ipsilateral lateral lemniscus. The remainder of the intermediate stria continue to the opposite side and become part of the contralateral lateral lemniscus.

Dorsal stria fibers ascend in the contralateral lateral lemniscus. Scattered neurons along the course of the lateral lemniscus constitute the nucleus of that tract and some ascending fibers synapse upon these cells.

Axons arising from the nucleus of the lateral lemniscus ascend with the remainder of the lateral lemniscus.

The lateral lemniscus (relaying impulses from the contralateral ear and some from the ipsilateral ear) terminates in the inferior colliculus of the midbrain.

Axons from cells of the inferior colliculus decussate to synapse upon neurons in the contralateral inferior colliculus. Other axons of the inferior colliculus neurons continue the auditory pathway, via the brachium of the inferior colliculus, to synapse in the medial geniculate body (M.G.B.) of the thalamus.

Projections from the medial geniculate body constitute the auditory radiations, which course laterally beneath the lenticular nucleus (putamen and globus pallidus) to their cortical destination. Transverse temporal gyri of Heschl in the floor of the lateral fissure, and the superior surface (bordering the lateral fissure) of the superior temporal gyrus constitute the primary auditory cortex.

Because of the intermingling of auditory impulses occurring in the inferior colliculi, the primary auditory cortex receives input derived from both ears, but from the contralateral ear predominately. Some inferior colliculus neurons are involved in mediating, via tectospinal and tectobulbar pathways, reflex postural adjustments in response to sudden sounds, in addition to being part of the cortically directed auditory pathway.

NOTE: The auditory nerve contains some efferent neurons which arise from the region of the superior olivary nucleus and terminate upon the organ of Corti. These fibers constitute the efferent cochlear bundle, synapsing upon outer hair cells of the organ of Corti.

Tonotopic Representation in the Auditory System

Low Tones	High Tones	
Upper	Lower	coils of the organ of Corti
Ventral	Dorsal	cochlear nuclei
Lateral	Medial	portion of the medial geniculate body
Antero/Lateral	Postero/Medial	parts of the gyri of Heschl

NOTE: Central lesions of the auditory pathway result in bilateral deficits, though usually more pronounced in the contralateral ear.

              Peripheral lesions result in unilateral hearing deficits.

NOTE: Eighth nerve tumors may occur and frequently arise from the sheath of the vestibular division within the internal auditory canal. As this tumor expands and encroaches into the posterior cranial fossa, there is compression of both divisions of the eighth nerve (vestibular and cochlear) and also the facial nerve which travels in the same canal.

CRANIAL NERVE DEFICITS

OLFACTORY (I)

Conveys the sense of olfaction from receptors in the nasal mucosa via nerve fibers which pass through the cribriform plate of the ethmoid bone. Relayed by cells in the olfactory bulb and back to the medial and lateral olfactory cortices in the olfactory tract.

Anosmia (impaired sense of smell) may result from head injury with fracture of the cribriform plate. Meningiomas arising from the olfactory groove or gliomas of the frontal lobe may similarly compromise the olfactory tract. Olfactory hallucinations may precede seizure activity arising from the anterior temporal lobe.

OPTIC (II)

Receptors for vision in the retina transmit impulses through the optic nerve via the optic canal, chiasm, optic tract, lateral geniculate body and visual radiations to the visual cortex in the medial occipital lobe. Monocular visual impairment can ensue with optic nerve gliomas or other retrobulbar masses in the orbit. Pupil reflexes are impaired following stimulation of the affected eye.

Superiorly expanding pituitary tumors, circle of Willis aneurysms or other mass lesions in the suprasellar region may compromise optic nerve, chiasm or tract (see section on visual system). Unilateral lesions of the visual pathway posterior to the chiasm cause contralateral visual field deficits (temporal lobe gliomas and low parietal lobe gliomas for example). Pupil light reflexes are spared with lesions of the visual radiations.

OCCULOMOTOR (III)

Arises from the midbrain in the interpeduncular fossa passing between the proximal portions of the posterior cerebral and superior cerebellar arteries. Travels in the lateral wall of the cavernous sinus (in the company of the trochlear, trigeminal (V1 and V2) and abducens nerves) via the superior orbital fissure into the orbit. Somatic fibers innervate the superior, medial and inferior rectus muscles, the inferior oblique and levator palpebri superioris muscles, visceral parasympathetic fibers relay via the ciliary ganglion to supply the ciliary and pupillary constrictor muscles.

Oculomotor nerve compression (as by aneurysm of the posterior communicating artery or cavernous sinus pathology for example) impairs the pupil response in the affected eye (pupil dilated and non responsive), paralyses the extraoccular muscles supplied by the third nerve (eye turned outward) with diplopia and results in ptosis (drooping eyelid).

TROCHLEAR (IV)

From the dorsal midbrain passing around the cerebral peduncle this nerve travels to the orbit via the wall of the cavernous sinus and the superior orbital fissure to innervate the superior oblique muscle.

Compression of the nerve along its course causes vertical diplopia upon attempted downward gaze. The head may be tilted over to the side opposite to the lesion to achieve compensation for the double vision.

ABDUCENS VI

Arises from the pontomedullary junction and supplies the lateral rectus muscle, travelling via th cavernous sinus and the superior orbital fissure.

Lesions of this nerve may cause inward deviation of the affected eye, although generally the eye may appear normal with forward gaze. Attempted lateral gaze to the affected side causes lateral diplopia as the affected eye does not abduct.

TRIGEMINAL (V)

General sensory innervation of the anterior scalp and face, including cornea and conjunctiva of the eye; upper teeth and mucous membrane of upper part of the mouth; mucous membrane of the floor of the mouth, lower teeth and tongue. (See section on trigeminal nerve). Distribution is via the three divisions of the nerve from the trigeminal ganglion via the superior orbital fissure (VI. ophthalmic), foramen rotundum (V2. maxillary), and foramen ovale (V3. mandibular). The trigeminal nerve attaches to the brainstem at the midpons level. The motor division of the trigeminal travels via the foramen ovale to innervate the muscles of mastication (temporalis, masseter, medial and lateral pterygoids) and the tensor tympani and veli palatini.

Trigeminal sensory loss may follow brainstem lesions (see section on anatomic localisation), cerebello pontine angle tumors (in association with other cranial nerve deficits), or peripheral nerve damage to one or more division of the nerve. Motor division involvement causes difficulty in chewing.

Facial pain may have many causes and may arise from sinusitis, teeth, temporomandibular joint problems, and ocular pain secondary to glaucoma to name some of the more obvious. Trigeminal neuralgia (tic douloureux) is a particularly excruciating condition with sharp periodic pain experienced over one or more divisions of the nerve which may be aggravated by stimulation of "trigger zones" on portions of the face. The cause of this condition is unclear many times, although in some cases irritation of the sensory root near its attachment to the brainstem caused by adjacent pulsatile arterial vessels may be the source of the problem.

FACIAL (VII)

The facial nerve contains motor fibers supplying muscles of facial expression, visceral motor fibers to the lacrimal, sublingual and submandibular glands, sensory fibers for taste (anterior 2/3 of the tongue), and fibers for general sensation for skin adjacent to the ear. The nerve emerges from the pontomedullary junction adjacent to the eighth nerve, courses through the internal auditory meatus, through the facial canal in the petrous temporal bone and exits via the stylomastoid foramen. Prior to passing through the stylomastoid foramen several branches are given off within the petrous bone. The greater petrosal nerve carries visceral efferent fibers which synapse in the pterygopalatine ganglion, from whence postganglionic parasympathetic fibers innervate the lacrimal gland. A small branch of the main motor component of the facial nerve departs to supply the stapedius muscle. The chorda tympani nerve leaves prior the stylomastoid foramen, carrying preganglionic parasympathetic fibers which synapse in the submandibular ganglion, postganglionics from there supplying the submandibular and sublingual salivary glands. The chorda tympani is also the course taken for special sensory taste fibers from the anterior 2/3 of the tongue. General sensory fibers from the skin round the ear enter through the stylomastoid foramen. Cell bodies of general sensory fibers and special sensory fibers (taste) are located in the geniculate ganglion within the depths of the petrous bone.

Brainstem lesions (e.g. infarcts) affecting the facial nucleus or peripheral nerve lesions (such as cerebellopontine angle tumors or acoustic tumors within the internal auditory canal) cause paralysis of all facial expresssion musculature on the affected side, with impaired eye closure. (Compare with corticobulbar lesions in the section on the pyramidal motor system). Lesions prior to the departure of visceral motor fibers will also affect lacrimation ("dry eye") and salivation. This may occur subsequent to skull base fractures. More distal lesions (parotid tumor for example, produce paralysis but spare lacrimation and salivation.

Bell's palsy (etiology unclear, but could be viral) affects the nerve in the facial canal or stylomastoid foramen, producing paralysis of mimetic musculature. Occasionally taste and lacrimation are also impaired, as is supply to the stapedius muscle.

AUDITORY (VIII)

Sound waves transmitted from the tympanic membrane transmitted through the ossicular chain, and converted into neural impulses in the cochlea by the organ of Corti, travel via the auditory division of the eighth nerve. Cell bodies of these sensory fibers are located in the spiral ganglion.

The vestibular division transmits impulses generated by receptors in the semicircular canal system responsive to angular and rotational movement and gravitational force (saccule and utricle). Cell bodies constitute the vestibular ganglion found within the petrous bone.

Both divisions traverse the internal auditory canal to attach to the brainstem at the pontomedullary junction. (See section on the auditory and vestibular system).

Hearing impairment may be due to problems related to sound conduction before the organ of Corti is reached (middle ear infection or trauma, for example). Similarly impairment may be due to involvement of the auditory nerve en route to the brainstem (e.g. acoustic nerve tumor or meningioma in the cerebellopontine angle). Acoustic tumors arising in the internal auditory meatus, as they expand will involve both divisions of the eighth nerve, then later, the facial nerve and occasionally the trigeminal and brainstem, as they expand into the posterior cranial fossa.

Tinnitus and increasing deafness indicate auditory division problems, vertigo and imbalance indicate vestibular malfunction.

GLOSSOPHARYNGEAL IX

This nerve carries taste fibers from the posterior 1/3 of the tongue (cell bodies in the petrosal (inferior) ganglion). This same ganglion also possesses cells which transmit information from the carotid body and sinus. Visceral motor fibers (parasympathetic) synapse in the otic ganglion and supply the parotid gland. General sensation from skin close to the ear is relayed by neurons of the superior ganglion. A motor component of the glossopharyngeal from the nucleus ambiguus supplies the stylopharyngeal muscle. The nerve exits the cranial cavity via the jugular foramen, in company with the vagus and spinal accessory nerves.

VAGUS (X)

This nerve, like the glossopharyngeal, contains numerous components. Taste sensation from around the epiglottis is relayed via neurons of the nodose (inferior) ganglion. Visceral motor fibers supply parasympathetic ganglia in the thorax and abdomen. Motor fibers (nucleus ambiguus) innervate muscles in the larynx, pharynx and soft palate. General sensation from around the ear, pharynx and larynx is carried by fibers having cell bodies in the jugular (superior) ganglion.

Lower brainstem lesions (e.g. infarcts) involving the glossopharyngeal and vagus nerves generally involve both simultaneously resulting in difficulty in swallowing, vocalization and hoarseness. Skull base tumors may involve both nerves as they exit the jugular foramen (and may also involve the spinal accessory nerve as it too exits the same foramen.

SPINAL ACCESSORY (XI)

This nerve, supplying the trapezius and sternocleidomastoid muscles, originates from the upper 4 or 5 cervical segments in the spinal cord, passes through the foramen magnum, and is joined by components from the nucleus ambiguus in the medulla. It leaves the cranial cavity via the jugular foramen. Pathology in the region of the foramen magnum may involve the spinal contribution. Skull base lesions near the jugular foramen may do the same (see above). Similarly, neck trauma may damage this nerve in its peripheral course. Drooping of the shoulder on the affected side will ensue (trapezius), and difficulty in turning the head to the side opposite the lesion (sternocleidomastoid).

HYPOGLOSSAL (XII)

Provides motor supply to the intrinsic tongue muscles, leaving the ventral medulla close to the pyramidal tract fibers and passing through the hypoglossal canal. Unilateral hypoglossal nerve lesions cause the tongue to deviate to the injured side upon attempted protrusion, with atrophy of that side of the tongue. Brainstem infarcts (see section on anatomic localization) may simultaneously involve the pyramidal tract giving rise to contralateral hemiparesis in addition to the hypoglossal signs.

THE DIENCEPHALON

The diencephalon, a large gray matter mass located at the rostral end of the brainstem, represents the highest subcortical level of neural interpretation. This important region has numerous complex functions which are involved in virtually all nervous system activity. This structure is partially divided into two halves by the midline presence of the third ventricle. The posterior limb of the internal capsule borders the diencephalon laterally, and the superior diencephalic surface forms part of the floor of the body of the lateral ventricle.

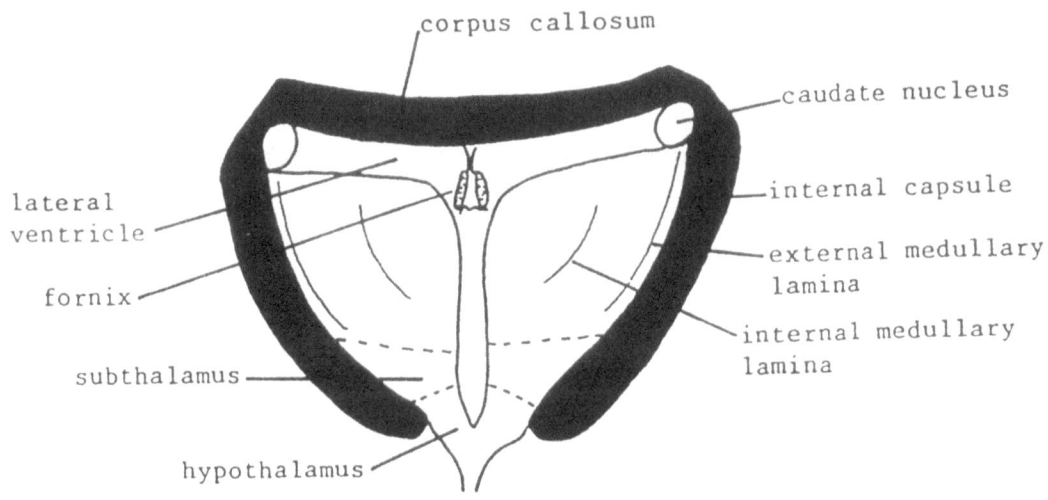

Coronal Section Through the Diencephalon

The entire structure may be divided into different subdivisions which are listed below.

DIVISIONS OF THE DIENCEPHALON

1. Epithalamus:

   Pineal organ, habenular nucleus and habenular commissure, posterior commissure, and stria medullaris thalami. This region is involved in olfactory and limbic system activity.

2. Hypothalamus:

   The ventral portion of diencephalon, largely concerned with autonomic, vegetative functions and relationship with the pituitary gland.

3. Subthalamus:

   Located inbetween the hypothalamus and the dorsal thalamus, this region includes the zona incerta and the subthalamic nucleus. It is concerned with extrapyramidal and basal ganglia activity.

4. <u>Thalamus</u> (dorsal thalamus):

By far the largest bulk of the diencephalon is comprised of the dorsal thalamus. It is important in relaying information to the cerebral cortex and in coordinating and interpreting many cortical functions.

5. <u>Metathalamus</u>:

This is actually part of the dorsal thalamus and refers to the medial and lateral geniculate bodies, which are of course special sensory relay nuclei.

The Dorsal Thalamus

The dorsal thalamus comprises numerous nuclear masses which fall generally into three principle categories.

1. Cortical integration nuclei concerned with complex integration of cerebral cortical activity.

2. Relay nuclei involved in projecting information from other regions of the nervous system towards various primary cortical regions.

3. A group of nuclei concerned with vegetative and autonomic activity relating to the functions of the hypothalamus and the brainstem reticular formation.

On coronal section, a sheet of white matter (internal medullary lamina) may be seen in the middle of the thalamic gray matter. Thalamic nuclei may also be grouped according to their relationship to the lamina, i.e., a dorsomedial group and a ventrolateral group.

The Ventrolateral Nuclei

The group of nuclei falling into the ventrolateral group may be conveniently divided into a ventral tier and a dorsal tier.

Ventral Tier Nuclei:	Anterior ventral Lateral ventral	. . Related to the basal ganglia and cerebellum
	Posterior ventral (medial) Posterior ventral (lateral) Medial geniculate body Lateral geniculate body	. . Specific sensory relay nuclei
Dorsal Tier Nuclei:	Pulvinar Lateral	Cortical integration nuclei, related to association areas of cerebral cortex

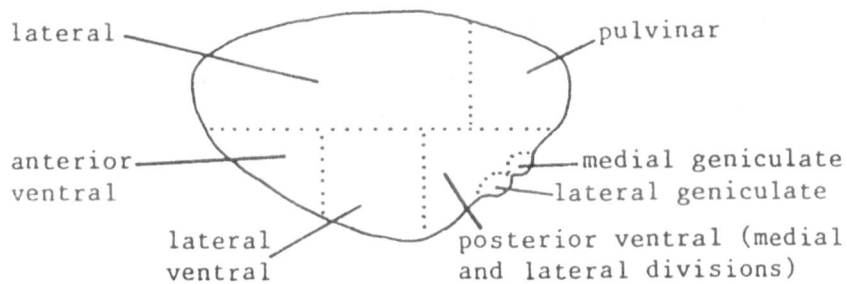

**Arrangement of the Ventrolateral Group
of Thalamic Nuclei**

PRINCIPLE NEURAL CONNECTIONS OF THE THALAMIC NUCLEI

I. Ventral Tier

A. Anterior Ventral Nucleus.

Afferents - Thalamic fasciculus from the basal ganglia. Receives a small additional input from the substantia nigra.

Efferents - to frontal lobe motor and premotor cortex primarily, but also some to the basal ganglia.

B. Lateral Ventral Nucleus.

Afferents - brachium conjunctivum from the cerebellum.

Efferents - to motor and premotor cortex of the frontal lobe.

C. Posterior Ventral Nucleus (lateral division).

Afferents - medial lemniscus and spinal lemniscus, relaying general sensation from the body below the head.

Efferents - to the postcentral gyrus (general sensory cortex) in the parietal lobe.

D. Posterior Ventral Nucleus (medial division).

Afferents - trigeminal lemniscus (general sensation from the head and face).

Efferents - to the inferior portion of the postcentral gyrus in the parietal lobe.

NOTE: The two divisions of the posterior ventral nucleus together constitute the largest sensory relay nucleus of the thalamus.

E. Medial geniculate body

   Afferents - brachium of the inferior colliculus

   Efferents - auditory radiations (sublenticular) to the transverse temporal gyri
   of Heschl in the temporal lobe.

F. Lateral Geniculate Body

   Afferents - optic tract

   Efferents - visual radiations (retrolenticular) to the primary visual cortex of
   the occipital lobe (calcarine cortex - cuneus and lingual cortex).

II. Dorsal Tier

A. Pulvinar.

   Reciprocal (both afferent and efferent) connections with the association cortex
   of the posterior parietal, lateral occipital and posterior temporal lobes.

B. Lateral Nucleus.

   Reciprocal connections with the parietal lobe.

Connections of the Nuclei lying Medial to the Internal Medullary Lamina

A. Dorsomedial Nucleus

   Afferents - from the hypothalamus and from the cortex of the prefrontal lobe
   (the anterior portions of the frontal convolutions).

   Efferents - to cortex of the prefrontal lobe

B. Midline Nucleus

   Consists of a thin layer of gray matter adjacent to the wall of the third
   ventricle (including the massa intermedia - if present). It is concerned with
   vegetative functions and possesses reciprocal connections with the hypothalamus.
   (NOTE: Approximately 70% of brains possess a massa intermedia.)

The Medullary Laminae.

These structures which are comprised of white matter, have a certain amount of gray
matter associated with them. The internal medullary lamina anteriorly encircles the
anterior nucleus, and posteriorly the centromedian nucleus.

Anterior Nucleus.

Afferents - mammillothalamic tract from the mammillary body of the hypothalamus.

Efferents - to the gyrus cinguli of the limbic lobe.

Centromedian

Possesses abundant connections with most of the other nuclei of the dorsal thalamus and also connections with the basal ganglia. Has a role in general integration of thalamic activity.

Intralaminar Nuclei

In addition to the anterior and centromedian nuclei, the internal medullary lamina possesses small scattered clusters of cells embedded in the white matter, which are related to most other thalamic areas. The brainstem reticular formation projects to these nuclei, and there is also some input derived from the spinothalamic tracts.

Reticular Nuclei

These are small patches of gray matter found enmeshed in the fibers of the <u>external</u> medullary lamina, possessing diffuse connections with the brainstem reticular formation. Efferent projections are to most areas of the cerebral cortex. Functionally these nuclei are related to the reticular activating system.

NOTE: Lesions involving the dorsal thalamus such as vascular infarcts or tumors for example, may produce a wide range of neurological impairment. Included in this range could be sensory deficits due to involvement of sensory relay nuclei, motor problems if involving the lateral ventral or anterior ventral nuclei, intellectual deterioration (pulvinar or lateral nucleus) and so on. Furthermore, space occupying masses encroaching upon the third ventricle may seriously impede cerebrospinal fluid passage, leading to increasing intracranial pressure and its accompanying serious complications.

HYPOTHALAMUS

The hypothalamus is the most inferior portion of the diencepalon located at the base of the brain, immediately superior to the sella turcica. The pituitary gland in the sella is attached to the inferior aspect of the hypothalamus by the pituitary stalk. The superior surface of the hypothalamus forms the inferior limit of the third ventricle, the infundibular recess of which extends a short distance into the infundibular portion of the pituitary stalk where it attaches to the hypothalamus.

The hypothalamus is a rather complex structure containing a number of nuclear groups. It also receives abundant input from numerous sources commensurate with its role of integrator of various visceral, metabolic and endocrine functions.

A convenient way to consider the hypothalamus is to divide it into three regions which are disposed rostrocaudally as follows:

1. Supraoptic area, most anterior and lying superior to the optic chiasm.
2. Tuberal area, located in the region of attachment of the pituitary stalk.
3. Mammillary area, the most posterior part of the hypothalamus.

Additionally, the hypothalamus may be divided into medial and lateral portions by the fibers of the fornix which traverse the hypothalamus en route to their termination in the mammillary bodies.

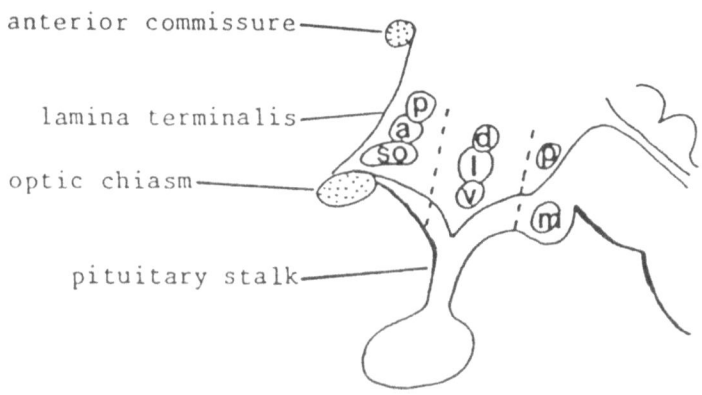

Position of the Hypothalamic Nuclei

1. Supraoptic Area

Nuclei   1) Paraventricular Nucleus   The neurons of the first two of these nuclei are well defined and contain cytoplasmic inclusions of neurosecretory material.

2) Supraoptic Nucleus

3) Anterior Nucleus

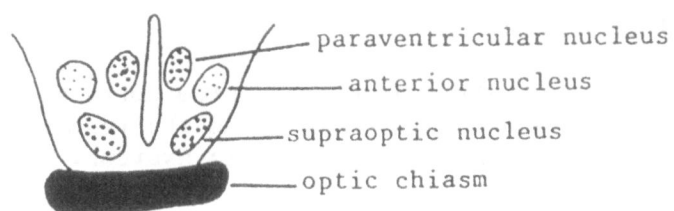

Coronal Section of the Supraoptic Area

2. <u>Tuberal</u> <u>Area</u>

      Nuclei   1) Ventromedial nucleus⎫ . . Medial to the fornix
                  2) Dorsomedial Nucleus ⎭
                  3) Lateral Nucleus . . . . . Lateral to the fornix

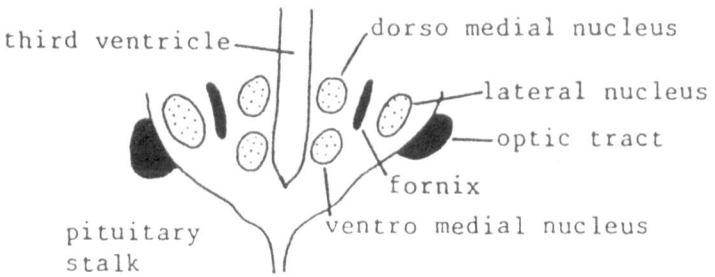

Coronal Section of the Tuberal Area

3. <u>Mammillary</u> <u>Area</u>

      Nuclei   1) Posterior Nucleus
                  2) Nucleus of the mammillary body

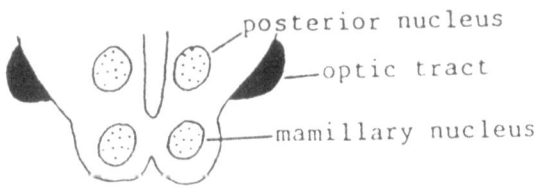

Coronal Section of the Mammillary Area

MAIN AFFERENT CONNECTIONS OF THE HYPOTHALAMUS

1. **Medial forebrain bundle.**

    An afferent and efferent fiber system connecting most of the hypothalamic nuclei with the septal area rostrally, and the tegmentum of the midbrain caudally. Related to the olfactory and limbic systems.

2. **Fornix.**

    Provides input from the hippocampal formation in the temporal horn of the lateral ventricle. As the fornix reaches the interventricular foramen of Monro, some fibers pass anterior to the anterior commissure, although the bulk of them pass posteriorly. Precommissural termination is in the supraoptic area and septal region. Postcommissural termination is in the mammillary body nucleus. This input is also related to the limbic system.

3. **Pallidohypothalamic Fibers.**

    Basal ganglia activity provides input into the hypothalamus via these fibers which arise from the globus pallidus.

4. **Stria Terminalis.**

    This system provides another connection between the hypothalamus and the olfactory and limbic system. From the medial portion of the amygdaloid nucleus in the temporal lobe to the anterior hypothalamus and septal region. It follows the course of the lateral ventricle adjacent to the medial border of the caudate nucleus.

5. **Visceral and Somatic Sensory Input.**

    From the lemniscal systems and the reticular formation.

6. **Periventricular Fibers.**

    Provide input derived from the midline and dorsomedial nuclei of the dorsal thalamus.

MAIN EFFERENT PATHWAYS

1. **Medial Forebrain Bundle** (see above).

2. **Mammillothalamic tract.**

    From the mammillary bodies to the anterior nucleus of the dorsal thalamus. This forms part of the circuit of Papez in the limbic system.

3. **Supraopticohypophyseal Tract.**

    From the supraoptic and paraventricular nuclei, down the stalk of the pituitary to the neural lobe of the pituitary. (Neurosecretory, transmitting antidiuretic hormone and oxytosin).

4. <u>Dorsal Longitudinal Fasciculus</u> (of Schutz).

   Travels in the ventral portion of the periaqueductal grey matter, to the brainstem reticular formation and autonomic nuclei.

5. <u>Tuberohypophyseal Fibers.</u>

   From the median eminence of the tuberal region to the vascular hypophyseal portal system in the pituitary stalk. These fibers convey hypothalamic releasing factors to the capillary bed of the portal system which in turn communicates with the sinusoids of the anterior pituitary. They function to regulate release of anterior pituitary hormones.

NOTE: The infundibulum (ventral portion of the hypothalamus where the stalk of the pituitary gland is attached) contains a high concentration of acetylcholine, and also some dopamine. Dopamine content is increased during pregnancy and lactation.

HYPOTHALAMIC FUNCTIONS

NUMEROUS ! ! ! ! ! ! ! ! ! ! !

One important function is in helping to regulate the activity of the autonomic nervous system. In general, it can be stated that the anterior hypothalamus regulates parasympathetic responses, whereas the posterior portion regulates sympathetic activity.

Some hypothalamic functions more specifically.

1. <u>Temperature Regulation.</u>

   The anterior portion of the hypothalamus responds to changes in temperature of the blood which supplies it.

   Anterior hypothalamus – stimulation activates mechanisms resulting in heat loss, sweating and vasodilation.

   Posterior hypothalamus – stimulation activates sympathetic mechanisms resulting in heat conservation such as peripheral vasoconstriction. Additionally somatic mechanisms are stimulated resulting in somatic muscle activity (e.g. shivering) which helps to increase body temperature.

2. <u>Water Balance Regulation.</u>

   Some anterior hypothalamic neurons are sensitive to osmotic changes in blood plasma, influencing the release of the antidiuretic hormone (A.D.H.). A.D.H. acts upon renal tubules to regulate the amount of water eliminated in the urine.

NOTE: Diabetes insipidus may be evident following damage to the supraoptic region or the pituitary stalk.

3. <u>Feeding Regulation and appetite.</u>

Ventromedial nucleus - satiation (lesions produce hyperphagia - excessive eating and and obesity).

Lateral nucleus - feeding (lesions produce hypophagia - reduced eating).

4. <u>Sleep-Wakefulness Regulation.</u>

Mediated via connections with the ascending reticular activating system. Lesions in the dorsolateral mammillary region may produce somnolence.

5. <u>Hormonal Regulation.</u>

Release of pituitary hormones regulating many bodily functions is controlled by releasing factors produced in the tuberal hypothalamic area and delivered via the hypophyseal portal vessels to the anterior pituitary.

6. <u>Regulation of Emotions.</u>

Mediated by numerous interconnections with the limbic system, involving hypothalamic integration of limbic system effector mechanisms. Lesions in the ventromedial nucleus may produce rage responses or hyperirritability.

NOTE: Sexual development may be altered in the presence of tumors in the region of the hypothalamus. This may lead to either precocious puberty, or conversely to hypogonadism and failure of development of secondary sexual characteristics.

Because of the multiplicity of functions with which the hypothalamus is involved, a wide variety of symptoms may result from hypothalamic malfunction. Sexual development may be altered (see above). Endocrine disturbances and changes in metabolic activity may ensue. Autonomic imbalance and emotional disturbances may also occur subsequent to hypothalamic disturbance. Additionally, tumors in this region may also compress the third ventricle and interfere with C.S.F. circulation and may produce hydrocephalus.

AUTONOMIC NERVOUS SYSTEM

The autonomic nervous system is comprised of two portions referred to as the sympathetic division and the parasympathetic division. Most structures which possess autonomic innervation receive input from both divisions of the system.

The autonomic system innervates peripheral effectors (smooth muscle, cardiac muscle and glands) by means of a two neuron system linking the C.N.S. with the structure being innervated. A preganglionic neuron (cell body lies within the C.N.S.) synapses with a postganglionic neuron (cell body lies outside the C.N.S.), which then terminates upon the effector structure.

The two divisions of this system generally produce opposing results in regard to their effect upon target structures. For example, sympathetic activation may cause heart rate to increase, whereas the response of parasympathetic discharge to heart muscle is a reduction of the rate of contractions.

Overall regulation of activity of autonomically innervated structures is achieved by balancing the activity of both portions of this system. Within the C.N.S. the hypothalamus represents an important controlling and integrating center and modulates the activity of both divisions of the autonomic system.

SYMPATHETIC DIVISION (Thoracolumbar system)

Preganglionic cells are located in the intermediolateral cell column of the spinal cord from the first thoracic segment down to the second lumbar segment.

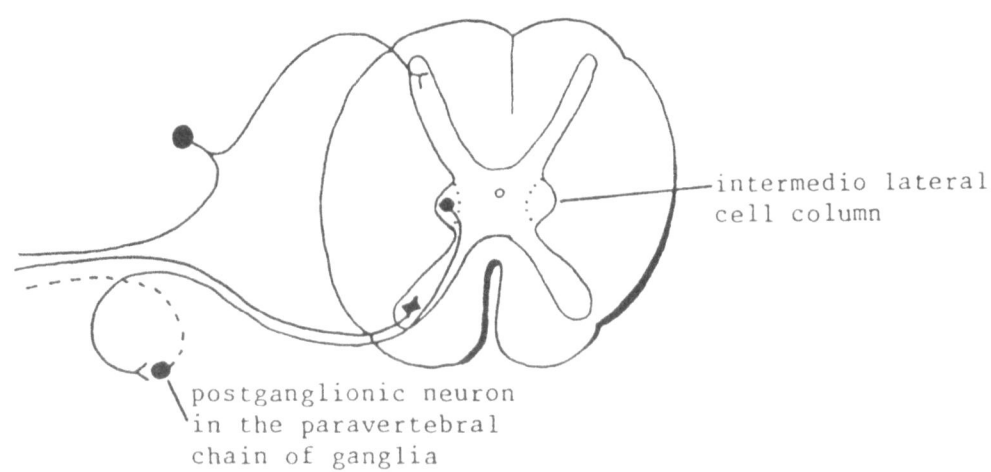

intermedio lateral
cell column

postganglionic neuron
in the paravertebral
chain of ganglia

Cord Section in the Thoracic Region showing the
Origin of the Sympathetics

Axons leave with the ventral roots of spinal nerves and travel via white rami communicantes to the paravertebral sympathetic ganglion chain. This bilateral series of sympathetic ganglia and the fibers which interconnect them rostrocaudally, is located on the posterior body wall adjacent to the bodies of the spinal vertebrae. At the level of entry into the sympathetic chain, preganglionic axons may do several things:

(1) synapse on postganglionic cells in chain ganglia at that level. Post ganglionic fibers then rejoin spinal nerves via the gray ramus and are distributed along the course of that nerve.
(2) synapse with postganglionics in chain ganglia at higher or lower levels in the chain (e.g., cervical sympathetic ganglia).
(3) pass from the chain to prevertebral ganglia (via splanchnic nerves) to synapse on cells of the prevertebral ganglia.

Prevertebral Ganglia.

These sympathetic ganglia are located at the origin of the unpaired branches of the abdominal aorta. They are the Coeliac, Superior mesenteric and Inferior mesenteric ganglia, giving rise to postganglionic fibers which travel to their destination on the outside of branches of these blood vessels.

PARASYMPATHETIC DIVISION (Craniosacral system)

This portion of the autonomic system arises in part from the brainstem, and also from sacral levels of the spinal cord.

Cranial Portion.

Preganglionic fibers originate from brainstem autonomic nuclei associated with cranial nerves III (oculomotor), VII (facial), IX (glossopharyngeal), X (vagus).

Brainstem Autonomic Nuclei.

1. Edinger Westphal Nucleus - upper midbrain.

   Preganglionics travel with the oculomotor nerve to the ciliary ganglion in the orbit. Postganglionic fibers are distributed to the ciliary muscle and pupil constrictor muscles.

2. Superior Salivatory Nucleus - Pontomedullary junction.

   Preganglionics are carried with the facial nerve synapsing in the pterygopalatine (sphenopalatine) ganglion and the submandibular ganglion. Postganglionics innervate the lacrimal gland in the orbit, and the submandibular and sublingual salivary glands.

3. Inferior Salivatory Nucleus - Medulla.

   Preganglionics carried with the glossopharyngeal nerve to the otic ganglion. Postganglionics innervate the parotid salivary gland.

4. Dorsal Vagal Motor Nucleus - medulla.

The vagus nerve distributes preganglionic fibers to diffuse terminal ganglia located in the walls of thoracic and abdominal viscera. Cardiac, pulmonary and myenteric plexuses are composed of the termination of these preganglionic fibers, the postganglionic neurons scattered among them, and postganglionic fibers.

In the abdomen, the vagus distributes to visceral structures approximately down to the level of the splenic flexure of the colon.

Sacral Portion

Preganglionic cells are located in the intermediate gray matter of the sacral segments of the spinal cord (S2,3,4). These neurons give rise to pelvic splanchnic nerves which are distributed to the pelvic viscera. Postganglionic cells lie in the visceral walls.

Sympathetic Functions

Generally concerned with arousing the body for immediate action related to self preservation, usually with a widespread effect. Dilation of pupils, increase in heart rate and respiration rate, increase in blood flow to cardiac and somatic muscles, reduction in immediate "non vital" functions (e.g., gut motility) would all be examples of response to sympathetic activation. Sympathetic function is controlled by the posterior portion of the hypothalamus.

Parasympathetic Functions

Generally related to anabolic activities and energy conservation. Constriction of pupils, decrease in heart rate, release of digestive secretions, and peristalsis of the gut illustrate some parasympathetic responses. These activities are under the regulation of the anterior part of the hypothalamus.

OLFACTORY AND LIMBIC SYSTEMS

These two systems are reviewed together as they are closely related to each other, and a number of structures are common to both systems. Olfactory sensation serves an obvious function in providing us with information regarding this aspect of the surroundings. However, the sense of olfaction may arouse a wide variety of emotional responses which are largely governed by limbic system activity. Thus the limbic system receives a considerable input from olfactory sources in addition to input derived from many other areas.

OLFACTORY SYSTEM

Receptors are found in the upper portion of the nasal cavity, and comprise specialized bipolar neurons embedded in the olfactory mucosa. Hair-like processes extend out from the mucosal surface and these respond to odor. Central processes form the olfactory fila and enter the cranial cavity by passing through foramina in the cribriform plate of the ethmoid bone. These primary fibers end here, by synapsing on dendrites of multipolar mitral neurons in the olfactory bulb, which lies in the olfactory groove on the under surface of the frontal lobe.

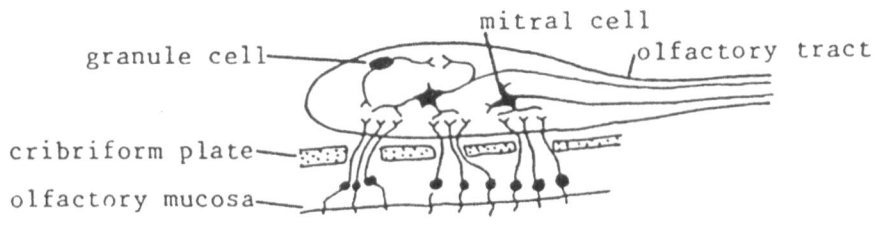

Olfactory Bulb and Nerves.

Axons of the mitral cells form the substance of the olfactory tract which passes caudally to attach to the brain.

Along the course of the olfactory tract are neurons which constitute the anterior olfactory nucleus. Some of the mitral cell axons synapse upon the cells of the anterior olfactory nucleus. Axons of the neurons of this nucleus may pass into the contralateral olfactory tract via the anterior commissure, to synapse on small granule cells in the olfactory bulb, helping to reinforce and enhance olfactory input.

Most axons of the olfactory tract attach to the base of the brain in the region of the anterior perforated substance close to the optic chiasm and the lamina terminalis.

The tract divides at its point of attachment into two well defined olfactory stria designated medial and lateral. A less well defined intermediate stria may be seen entering the anterior perforated substance.

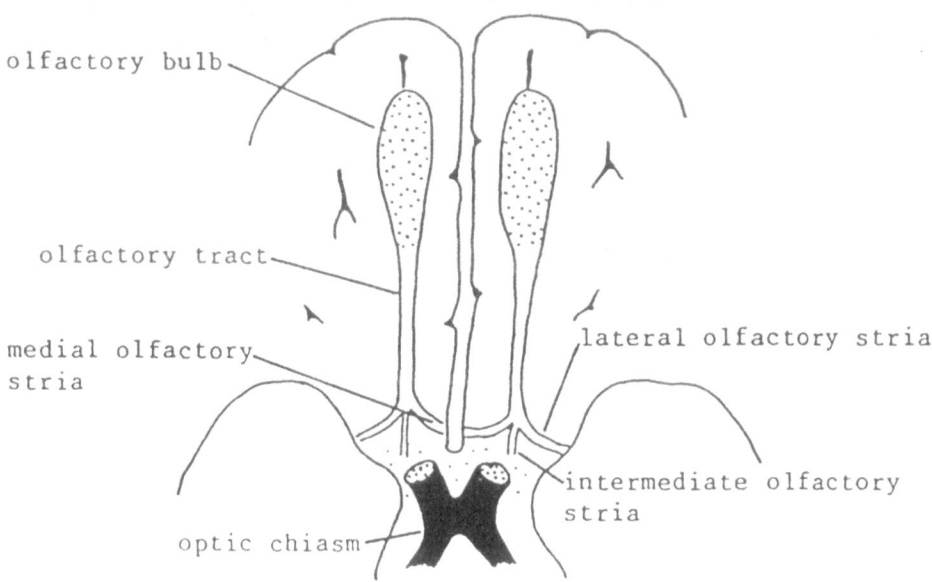

Under Surface of the Brain Showing Olfactory Connections

The lateral olfactory stria terminates in the prepyriform cortex which lies immediately in front of the uncus on the parahippocampal gyrus of the temporal lobe. This region constitutes the primary olfactory cortex.

The medial olfactory stria terminates upon cortex of the medial side of the frontal lobe in the subcallosal gyrus (in the septal area).

NOTE: Olfaction is the only sensory modality which reaches the cerebral cortex without first being relayed via a thalamic nucleus.

Much of the activity generated by olfactory sensation is directed into paths which are concerned with providing reflex responses to this input.

Reflex Pathways Used by Olfactory Input.

1. Medial Forebrain Bundle - from the medial olfactory cortex in the septal area, caudally to the hypothalamus and midbrain tegmentum.

2. Stria Medullaris Thalami - from the septal region to the habenular nucleus in the epithalamus. The habenulo interpeduncular tract arising from the habenular nucleus makes connection with the interpeduncular region of the midbrain tegmentum.

3. Stria Terminalis - from the amygdaloid nucleus (activated by neurons in the lateral olfactory cortex) to the septal region, preoptic area and anterior hypothalamus. This fiber tract courses along the border of the lateral ventricle, adjacent to the caudate nucleus.

By these pathways, olfactory input may result in reflex responses, many of which result in autonomic activity in visceral areas for control of salivation, mucous secretion, etc.

LIMBIC SYSTEM

Two terms are used in describing this apparatus, the limbic lobe, and the limbic system which includes the limbic lobe and other related structures.

The limbic lobe consists of areas of gray matter encircling the diencephalon and includes the following:

        Gyrus cinguli
        Parahippocampal gyrus
        Dentate gyrus
        Indusium griseum (hippocampal rudiments lying on top of the
                          corpus callosum)
        Hippocampus

The limbic system in addition to the above, includes further gray matter regions as follows:

        Septal area
        Orbital gyri
        Anterior thalamic nucleus
        Hypothalamus
        Amygdaloid nucleus
        Epithalamus
        Midbrain tegmentum
        Insula cortex

The various parts of the limbic system are interconnected via numerous fiber tracts and pathways. Some of these pathways are shared with the olfactory system.

Fiber Tracts involved in Limbic System Function.

Medial forebrain bundle
Stria terminalis
Stria medullaris thalami
Habenulo interpeduncular tract
Fornix
Mammillothalamic tract
Anterior commissure
Cingulum
Uncinate fasciculus

Limbic System Function.

Regulation of emotional tone and behavior in general summarizes the result of limbic system activity. By way of the numerous connections with the hypothalamus, the limbic system is intimately related to visceral and autonomic areas, by means of which much of limbic system activity is expressed. The limbic system receives widespread input from numerous cortical regions to enable it in the complexities of emotional regulation.

NOTE: Lesions in the system may produce a variety of changes in behavioral activity.

Medial temporal lobe damage involving the uncus may give rise to olfactory hallucinations.

Hippocampal involvement may lead to memory impairment, particularly in the processing of short term memory. Bilateral damage produces particularly severe deficits in this regard. Long term memory stored in regions other than the hippocampal formation, is not generally affected. Memory impairment is also seen with mammillary body damage as seen in Korsakoff's syndrome.

Bilateral anterior temporal lobe damage involving the amygdaloid nuclei and adjacent cortex may result in striking behavioral changes. Lack of apparent awareness of potentially threatening situations, an unusual degree of apparent curiosity in surrounding objects (many of which are examined orally), indiscriminate hypersexuality (particularly in males) are all components of what has been called the Kluver-Bucy syndrome resulting from anterior temporal lobe damage. Some of these symptoms may be seen subsequent to anterior temporal lobe contusions following head injury.

CEREBRUM

The cerebral hemispheres consist of a surface layer of cortical gray matter, a central core of subcortical white matter, and several subcortical gray matter masses embedded in the deeper portion of the white matter (the basal ganglia). Cerebropinal fluid filled ventricular cavities (the lateral ventricles) occupy the central portion of each hemisphere.

The 2 - 4 mm. thick layer of cerebral cortex possesses a complex cytological arrangement and has been estimated to contain somewhere in the region of 14 - 15 billion neurons. The interrelationship which can exist between these cells and other areas of the nervous system permit the enormous diversity of behavior of which humans are capable. The so called "higher functions" of cerebration such as ideation, intellectual ability, language and symbolism for example, are human characteristics which are only possible subsequent to the development of the enormous amount of neocortex.

It is possible, to a certain degree, to assign certain functions to regions or lobes of the cerebral hemispheres, although many of these functions are by no means solely confined to precisely defined areas.

Major lobes of the cerebrum are defined by their relationship to certain fissures or deep sulci found on the cerebral surface. The central sulcus (of Rolando) and the lateral fissure (of Sylvius) on the lateral surface of the hemisphere; the circular sulcus in the depths of the lateral fissure; the parietooccipital sulcus and cingulate sulcus on the medial side of the hemisphere, define some of the boundaries.

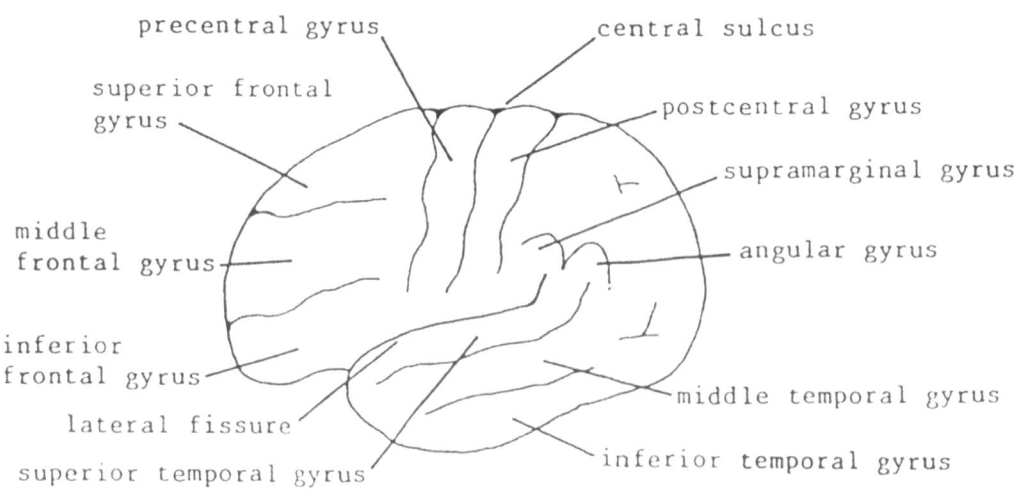

Lateral View of the Left Cerebral Hemisphere

HENCE:

Frontal lobe		Superior frontal gyrus
		Middle frontal gyrus
		Inferior frontal gyrus
		Precentral gyrus
		Orbital gyri
		Straight gyrus

Temporal lobe

Transverse temporal gyri (of Heschl)
Superior temporal gyrus
Middle temporal gyrus
Inferior temporal gyrus
Fusiform gyrus
Parahippocampal gyrus

Parietal lobe

Supramarginal gyrus
Angular gyrus
Superior parietal lobule
Precuneus
Postcentral gyrus

Occipital lobe

Lateral occipital gyri
Cuneus
Lingual gyrus

Insula   Bounded by the circular sulcus

Cingulate gyrus   Bounded by the cingulate sulcus

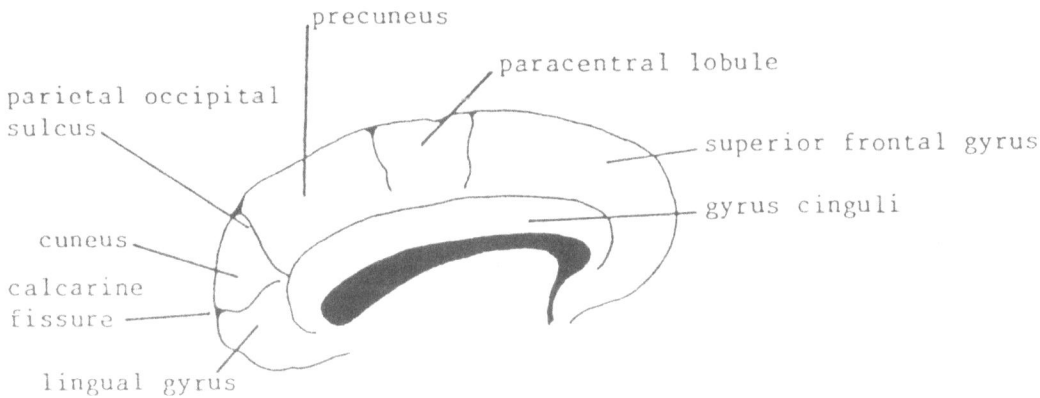

Medial View of the Left Hemisphere

CORTICAL HISTOLOGY

Numerous different cell types more or less disposed in a layered arrangement. Most common cell types found in the cortex are pyramidal neurons (of several sizes) and stellate neurons. Other cells found, but in lesser numbers, are horizontal cells of Cajal, cells of Martinotti, and irregularly shaped polymorph neurons. In addition to these neurons, numerous glial cells are found, astrocytes particularly.

Six layers may be described (numbered 1-6 from superficial to deep):

1. Molecular layer - horizontal cells of Cajal and dendrites of neurons located in the deeper cortical layers.

2. Outer granular layer - small pyramidal cells.

3. Outer pyramidal layer - medium sized pyramidal cells.

4. Inner granular layer - stellate cells.

5. Inner pyramidal layer - large pyramidal cells.

6. Polymorph layer - irregularly shaped neurons.

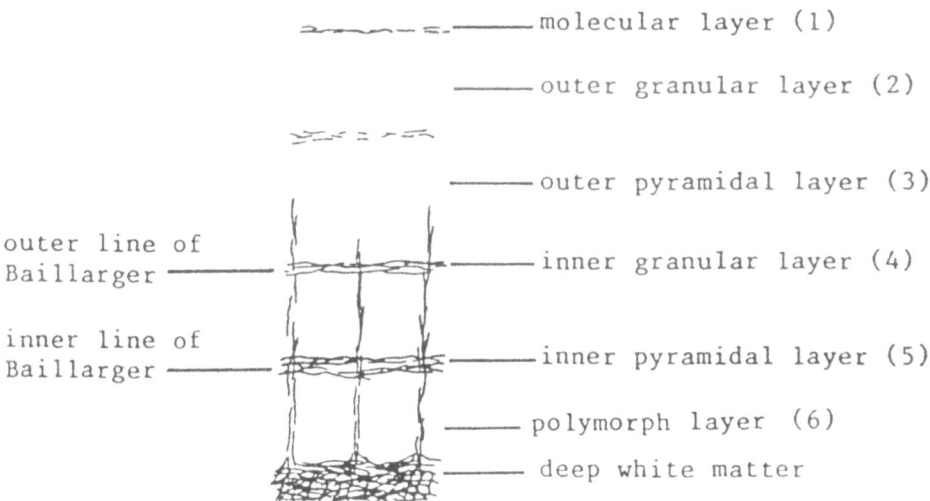

Section of Cerebral Cortex Following a Fiber Stain

Pyramidal cells
     These possess an apical dendrite which is directed towards the outer surface of the cortex. The axon enters deeper layers of the cortex or extends into the deep white matter.

Stellate cells
     Have short axons and dendrites which are confined to the cortex.

Martinotti cells
     Found in most layers, they have short dendrites and axons extending to the cortical surface.

Horizontal cells of Cajal
     Axons and dendrites course parallel to the cortical surface in the molecular layer.

Polymorph cells
>Dendrites extend towards more superficial layers, axons into the deep white matter.

Sensory Cortex Layer 4 (inner granular layer - stellate cells) is particularly prominent and pyramidal cell layers are reduced.

Motor Cortex Layers 3 and 5 (outer pyramidal and inner pyramidal layers) are prominent.

Most afferent fibers entering the cortex terminate in the inner granular layer (Layer 4) or the inner pyramidal layer (Layer 5). With sections stained to show nerve fibers rather than cell bodies, these terminations are demonstrated by fiber bands (outer and inner lines of Baillarger) running parallel to the cortical surface.

Cerebral White Matter.

Deep white matter of the hemispheres is composed of neuronal axons and numerous glial cells. The axons may be considered to be of three basic functional types.

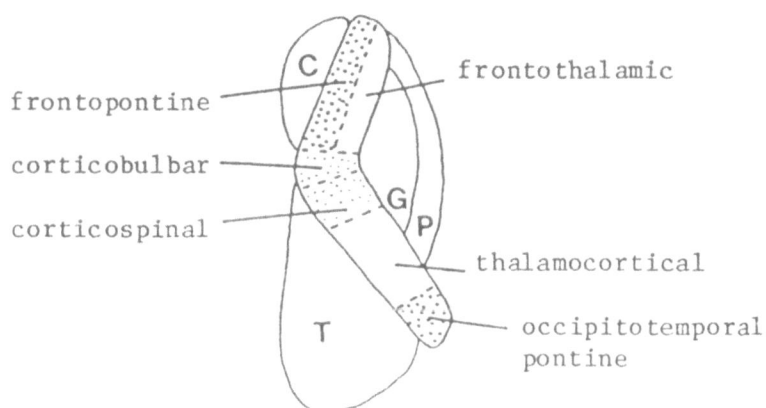

Horizontal Section of the Internal Capsule

1. Projection Fibers.

Axons connecting areas of cerebral cortex with other regions of the neuraxis, either brainstem, diencephalon or cord (e.g., corticospinal fibers). Also projection fibers include axons arising from lower areas (for example, the diencephalic nuclei) and terminating in the cerebral cortex. Most of these fibers are also seen in the internal capsule.

2. Association Fibers.

Axons connecting a specific region of cerebral cortex with other cortical regions in the same hemisphere.

3. Commissural fibers.

Axons connecting one cortical region with another cortical region in the opposite cerebral hemisphere. The corpus callosum consists of the largest bundle of commissural fibers, crossing the midline between the two hemispheres.

Association fibers may be relatively short and superficially located in the substance of the white matter. They may pass from one cortical gyrus, below a cerebral sulcus, to the adjacent cortical gyrus.

Other association fibers may be more deeply located in the white matter, and interconnect widely separated cortical regions (e.g., superior longitudinal fasciculus - connects frontal cortex with parietal and temporal cortices).

Commissural fibers pass from one hemisphere to the other, via the Corpus Callosum. The minor forceps, from one frontal lobe to the other frontal lobe, crosses in the genu of the corpus callosum. The major forceps, connecting both occipital lobes travels in the splenium of the corpus callosum.

FUNCTIONAL REGIONS OF CEREBRAL CORTEX

Cortical regions may be described as either primary or secondary regions. Primary regions are directly concerned with motor or sensory function. Secondary areas, to which the term association cortex can be applied, handle more complex functions.

Primary Cortex		Brodman's area
Primary Motor area	Precentral gyrus	4 & 6
Supplementary Motor area	Cortex of frontal gyri immediately anterior to the precentral gyrus	8
General sensory	Postcentral gyrus	3, 1, 2
Primary visual	Cuneus and lingual gyrus adjacent to the calcarine sulcus	17
Visual association area	Remainder of the cuneus and lingual gyri, plus the lateral occipital gyri	18 & 19
Auditory	Transverse temporal gyri of Heschl	41 & 42
Sensory (receptive) Speech area (Wernicke)	Supramarginal gyrus and the posterior portion of the superior temporal gyrus	40
Motor Speech (Broca)	Caudal portion of the inferior frontal gyrus of left hemisphere in the majority of people	44

Association Cortex

Prefrontal Cortex.

This region also known as the prefrontal lobe, consists of the anterior two inches or so of the frontal convolutions. It is concerned with personality, relationship of the individual to others, planning for the future, and concern for the outcome of one's behavior.

Parietal, Occipital and Temporal association cortices

These cortices are concerned with complex integration and interpretation of the sensory functions subserved by adjacent primary areas (general sensation, vision and auditory).

Olfactory.

Subcallosal cortex on the medial side of the frontal lobe and cortex of the parahippocampal gyrus of the temporal lobe anterior to the uncus (the prepyriform cortex).

Taste.

Opercular portion of the postcentral gyrus and the adjacent insula cortex.

Memory.

The hippocampal formation in the temporal lobe, from which arises the fornix, is necessary for processing memory storage (short term memory). Subsequently long term memory is no longer retained in the hippocampal formation, but is thought to reside in more widely distributed cortical associational areas.

Cortical eye fields.

Frontal - caudal portion of the middle frontal gyrus
Occipital - Lateral occipital gyri.

Both these cortical eye fields exert an influence over movements of the eyes within the orbits.

NOTE: Lesions involving the cerebral cortex may obviously lead to
          impairment of the function subserved by the affected area.
          For example, lesions of Wernicke's area may cause a
          receptive (or sensory) aphasia. Lesions involving Broca's
          area in the frontal lobe lead to motor (or expressive)
          aphasia. Temporal lobe lesions affecting the hippocampus
          may cause memory deficits. Visual deficits result from
          impairment of the medial occipital lobe cortex.

          Most of the examples given above are more or less
          definitive. However diffuse neuronal malfunction or
          degeneration, may produce a gradual impairment of the higher

functions of cerebral activity with little noticeable
initial impairment of the more basic neurological functions.
Levels of consciousness are determined by the overall
functional state of the cerebrum. Reticular formation
projection to the cerebral cortex is a necessary part of
normal cortical activity. Comatose conditions in which the
brainstem is compromised have a more dire prognosis than
similar conditions in which the brainstem is still intact.

VASCULARITY OF THE CENTRAL NERVOUS SYSTEM

INTRACRANIAL ARTERIAL SUPPLY

Unlike other body tissues, the nervous system has a limited capacity to store some of the material essential for normal function, glucose and oxygen in particular. Compromise of the delivery of oxygenated and glucose carrying arterial blood to nervous tissue results in malfunction of the region deprived of nourishment, and ultimate tissue death if the situation prevails for a long enough time.

Normally the central nervous system is supplied with arterial blood from a number of sources, allowing the possibility of alternate routes of delivery of necessary nutrients in some circumstances.

Two pairs of arteries are normally responsible for arterial blood supply to the brain:

  1. Internal carotid arteries (anterior circulation).
  2. Vertebral arteries (posterior circulation).

Internal Carotids enter the cranial cavity via the carotid canal, traverse the cavernous sinus adjacent to the pituitary fossa, and finally become intracranial in the middle cranial fossa close to the anterior clinoid processes and the optic chiasm.

The ophthalmic artery is given off just as the carotid leaves the cavernous sinus.

The next branches of the internal carotid are successively the posterior communicating artery and the anterior choroidal artery.

The internal carotid artery terminates as two main vessels, the anterior cerebral and middle cerebral arteries.

Vertebral Arteries enter the cranial cavity via the foramen magnum and unite to form the single basilar artery, anterior to the base of the pons.

Each vertebral artery gives rise to a posterior inferior cerebellar artery and also a branch which unites with a similar vessel from the other vertebral to form a single anterior spinal artery, which passes caudally to supply the spinal cord.

Branches of the basilar: Anterior inferior cerebellar arteries, internal acoustic arteries, pontine perforating vessels, superior cerebellar arteries, and posterior cerebral arteries. (NOTE: The posterior cerebral arteries are the terminal branches of the basilar).

THUS: 3 pairs of arteries supplying the supratentorial brain:

Anterior cerebrals.
                          . . . From the internal carotids
Middle cerebrals

Posterior cerebrals . . . From the vertebrobasilar vessels.

3 pairs of vessels supplying the infratentorial brain
(cerebellum and brainstem):

Superior cerebellar

Anterior inferior cerebellar ·········· all from the
vertebrobasilar
vessels

Posterior inferior cerebellar

Anastomosing vessels provide communication between the anterior and posterior
circulations – two posterior communicating arteries (distal carotid to proximal
posterior cerebral) and one anterior communicating (between both anterior
cerebrals).

Hence the Arterial Anastomotic Circle of Willis

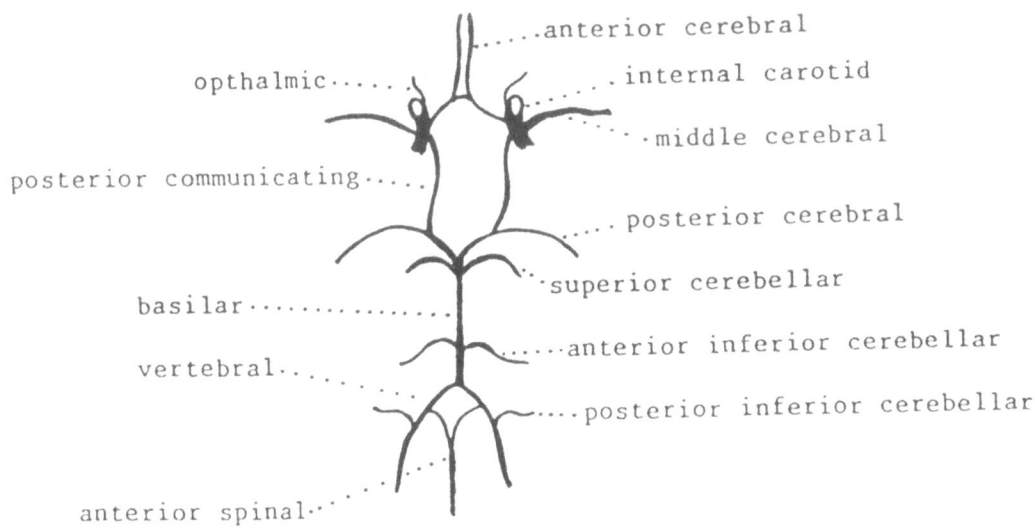

Major Arteries seen on the Base of the Brain

NOTE: Numerous small perforating vessels (many of which are not
           named) arise from the proximal portions of the cerebral
           vessels and directly enter the adjacent brain substance.
           One set of these vessels which are named, the
           lenticulostriate arteries, arise from the proximal portion
           of the middle cerebral artery to supply the lenticular
           nucleus (putamen and globus pallidus). These particular
           vessels are frequently the site involved in hypertensive
           intracerebral hemorrhage.

NOTE: The vessels of the Circle of Willis, and the proximal
           portions of the major cerebral vessels (especially at the
           points of branching, are favorite sites for the occurrence
           of saccular (Berry) aneurysms.

Regions Supplied by the Main Arteries

Anterior Cerebral.

Medial side of the cerebral hemispheres, above the corpus callosum, as far caudally as the parietooccipital fissure. Terminal branches extend over the interhemispheric fissure, to supply approximately 1 inch of the superior surface of the hemispheres.

Middle Cerebral.

Penetrating vessels through the anterior perforated substance to supply basal ganglia etc. Insula cortex, lateral aspect of the bulk of the cerebral hemispheres (frontal, parietal, occipital, and temporal lobes).

Posterior Cerebral.

Medial side of cerebral hemispheres caudal to the parietooccipital fissure, inferior aspect of the occipital and temporal lobes. Terminal branches overlap onto lateral aspect of the hemispheres by approximately 1 inch, similar to the terminals of the anterior cerebral.

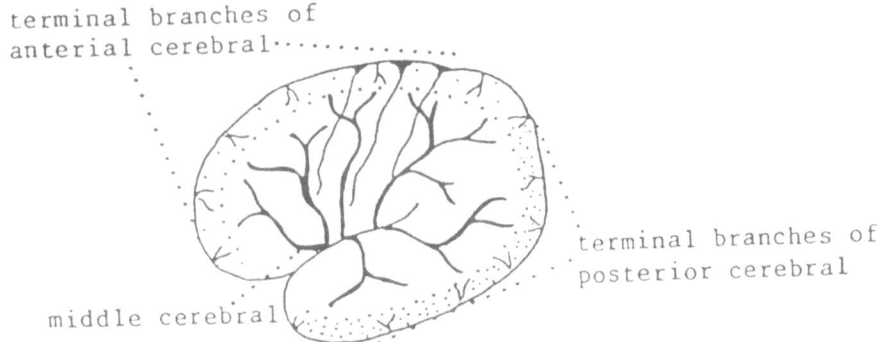

Cerebral Vessels from a Lateral View

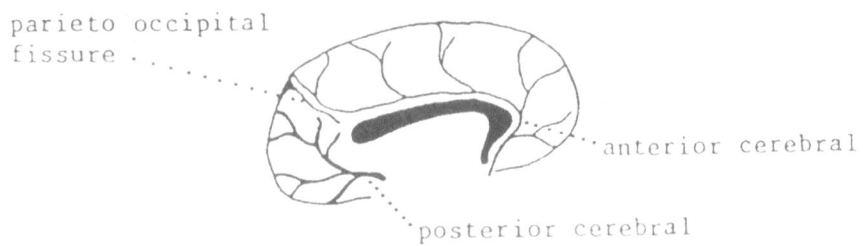

Medial View of Cerebral Vessels

NOTE: Structures deep to the surface are supplied by penetrating branches (central or ganglionic branches) of the surface vessel.

Cerebellar arteries are distributed over the cerebellar surface, generally in the
manner which their name implies. Penetrating branches also supply the adjacent
brainstem, by paramedian vessels entering from the ventral surface and
circumferential vessels supplying the lateral and dorsal regions.

INTRACRANIAL VENOUS DRAINAGE

Deep structures of the supratentorial compartment are drained via the paired
internal cerebral veins in the roof of the third ventricle. These unite to form the
short great vein of Galen just above the pineal organ. This drains into the
straight sinus at the junction of the falx and the tentorium. The straight sinus
empties into the confluence of sinuses (Torcula of Herophilus), then into the right
and left transverse sinuses, into the sigmoid sinuses and hence to the Jugular Bulb
and the Internal Jugular vein.

Venous blood which drains into the superficial cerebral veins over the lateral,
superior and medial portions of the hemispheres, mostly passes by way of veins
bridging the subarachnoid space to empty into the superior and inferior sagittal
sinuses.

Superior and inferior sagittal sinuses at the borders of the Falx, drain into the
confluent and straight sinuses, respectively. The inferior (basal) brain surface
veins drain into the superior and inferior petrosal sinuses, cavernous sinus,
spehenoparietal sinuses and the basilar venous plexus.

SPINAL CORD VASCULATURE

Paired posterior spinal arteries arising from the vertebrals and radicular arteries,
are found on the dorsal cord surface, near the entrance of dorsal nerve roots. A
single anterior spinal artery derived from the vertebrals, courses along the
anterior median fissure of the spinal cord. It receives contributions from
segmental radicular vessels along its course.

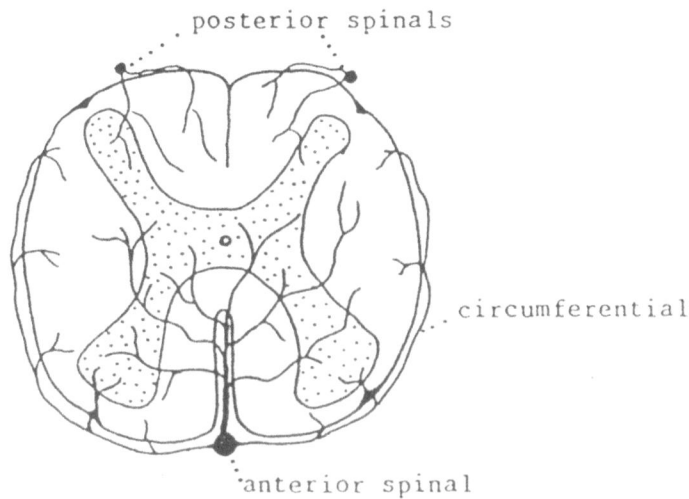

Arteries of the Spinal Cord

NOTE: Occlusion of the anterior spinal artery produces profound
         neurologic deficits due to its large territory of supply in
         the cord. The anterior spinal artery supplies approximately
         two-thirds of the cord cross section and includes the
         anterior horns of the grey matter and almost the entire
         lateral funiculus.

Spinal veins generally accompany the spinal arteries. They drain via radicular
veins into the epidural venous plexus, lying between the spinal dura and the walls
of the bony vertebral canal. Epidural veins communicate with the external vertebral
venous plexus round the vertebral bodies and hence to the systemic venous system.

NOTE: The epidural venous plexus may commonly be the site of
         development of metastases by seeding of primary neoplasms
         located in the abdominal or thoracic viscera. These
         secondary tumors may cause spinal cord compression and
         neurological signs, and this may be the first indication of
         the presence of the primary neoplasm.

CEREBROSPINAL FLUID CIRCULATION

From the site of production (choroid plexus and ventricular ependyma) through the
ventricular system into the subarachnoid space. The normal course of flow of C.S.F.
is successively through the lateral ventricles, interventricular foramina of Monro,
third ventricle, cerebral aqueduct (of Sylvius), fourth ventricle, lateral foramina
of Lushka and medial foramen of Magendie into the subarachnoid space surrounding the
lower brainstem. C.S.F. in the cranial subarachnoid space is normally continuous
with spinal subarachnoid C.S.F.

Absorption of C.S.F. is largely through arachnoid villi projecting into the superior
sagittal sinus. Some absorption also takes place at small villi close to the exit
of spinal nerves from the spinal dura.

Localized dilatations of the subarachnoid space are called cisterns.

Important ones:
         1. Lumbar cistern at caudal end of spinal cord
                              (vertebral levels L2 - S2).

NOTE: Diagnostic lumbar punctures are usually attempted at the
         L4-L5 or L3-L4 interspace.

         2. Cisterna magna (cerebellomedullary cistern) Between the medulla
                              and the cerebellum.
Other cisterns:
         Pontine cistern
         Interpeduncular cistern
         Chiasmatic cistern
         Quadrigeminal cistern
         Cisterna ambiens
         Retropulvinar cistern

NOTE: Impairment of the normal flow and drainage routes of C.S.F.
           leads to accumulation of fluid and hence hydrocephalus.
           Obstruction of ventricular flow leads to non-communicating
           hydrocephalus, in which C.S.F. in the subarachnoid space
           does not communicate with the C.S.F. in the blocked
           ventricular system. Communicating hydrocephalus is the
           condition which results if fluid passes freely from the
           ventricular system into the subarachnoid space, but is
           prevented from reaching sites of reabsorption.

Some examples of conditions leading to <u>non-communicating</u> hydrocephalus could be:

           Atresia or stenosis of the cerebral aqueduct.
           Failure of development of the foramina of Lushka and Magendie.
           Presence of a tumor obstructing the ventricular C.S.F. pathways.

Some possible causes of <u>communicating</u> hydrocephalus:

           Non-development or malfunctioning of arachnoid villi.
           Inflammation of the meninges (from blood in the subarachnoid
                                      space or infectious processes).

BLOOD BRAIN BARRIER

Unlike other regions of the body, most regions of the central nervous system are
characterized by possessing capillaries which permit selective passage of certain
blood borne substances, whilst precluding the entry of other substances from blood
to brain. This unique property of C.N.S. capillaries is largely due to the nature
of the cell membranes of capillary endothelial cells, and the presence of tight
occluding junctions between endothelial cells. Capillaries elsewhere in the body do
not display tight junctions and furthermore, many capillaries possess fenestrated
walls, both of which allow relatively free exchange of substances between blood and
extracellular fluid.

Although basically a protective mechanism, the blood brain barrier may sometimes
work to our disadvantage. For example, some medications cross the barrier with
extreme difficulty, or in some cases, not at all. In some cases, entry into the
brain may be brought about by administration of the medication directly into the
C.S.F.

The properties of the blood brain barrier mechanisms may be caused to change under
various conditions, included in which could be hypertension, changes in osmotic
pressure of cerebral blood supply, in the presence of some C.N.S. neoplasms, certain
infections or disease states.

Some regions within the brain are excluded from the protection of the blood brain
barrier. These regions include: The area postrema, subfornical organ, pineal
organ, median eminence, supraoptic crest, neurohypophysis and choroid plexus.

NOTE: Intracranial pressure may be altered by the vascular volume
           of the central nervous system. For example, if the C.S.F.

pressure is measured at the lumbar cistern, temporary compression of the internal jugular veins in the neck will cause an increase in C.S.F. pressure until the compression is released. The pressure will then be seen to return to the original level. The reason for this is jugular compression impairs the major intracranial venous drainage route leading to temporary increase in intracranial vascular volume, and hence increase in C.S.F. pressure (Queckenstedt test).

A similar response is seen if the abdominal pressure is increased (by holding one's breath for example). This is due to temporary partial compression of veins draining the spinal cord (Valsalva maneuver).

EARLY STAGES OF NERVOUS SYSTEM DEVELOPMENT

The nervous system develops embryologically from the neural plate portion of the ectoderm, which invaginates to form the neural groove. Further invagination of the neural groove brings about approximation of the dorsal lips of the groove thus forming the hollow neural tube.

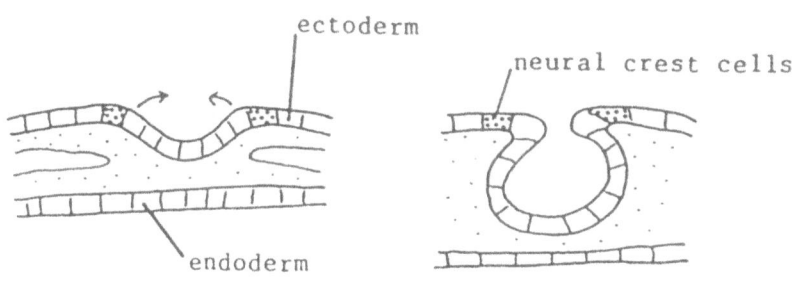

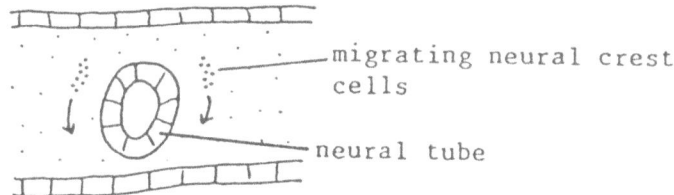

Development of Neural Tube

Neuroepithelial cells which constitute the neural tube divide to produce neuroblasts which eventually develop into neurons, and glioblasts which give rise to the supporting cells of the nervous system.

NOTE: Microglia are not derived from neuroectoderm but are of mesodermal origin, and enter the developing nervous system at a later stage of development along with developing blood vessels.

Neural crest cells derived from near the adjoined lips of the neural tube give rise to ganglia, sensory fibers of peripheral nerves, and autonomic nervous system ganglia and neurons.

The neural canal retains continuity with the amniotic cavity for a while via the anterior (at the cranial end) and posterior (caudal end) neuropores. The anterior neuropore closes first at about the 18-20 somite stage, later followed by the closure of the posterior neuropore at about the 26 somite stage.

As development progresses, the anterior end of the neural canal expands to form three primitive brain vesicles. At four weeks of development, these are the rhombencephalon (forming the hindbrain), the mesencephalon (forming the midbrain) and the <u>prosencephalon</u> (forming the forebrain), and by this stage, both neuropores should be completely closed.

Later in development (by six weeks) there are five vesicles which are derived from the primitive vesicles mentioned above:

Rhombencephalon derivatives

Myelencephalon - Medulla; glossopharyngeal, vagus, spinal accessory and hypoglossal nerves.

Metencephalon - Pons; trigeminal, abducens, facial, and auditory nerves.

Rhombic lip of the mesencephalon - Cerebellum.

Fourth ventricle.

Mesencephalon derivatives

Midbrain - Oculomotor and trochlear nerves.

Cerebral aqueduct.

Prosencephalon derivatives

Diencephalon (including the optic vesicles), optic nerves

The paired telencephalon - cerebral hemispheres (including the basal ganglia), olfactory nerves

Third ventricle.

The lateral ventricles.

As the embryo enlarges, the developing neural tube bends ventrally at two places. These bends are the <u>cervical</u> flexure at the junction of the spinal cord and the rhombencephalon, and the <u>cephalic</u> flexure, at the boundary between the rhombencephalon and the mesencephalon. These two flexures are present by the fourth week of development.

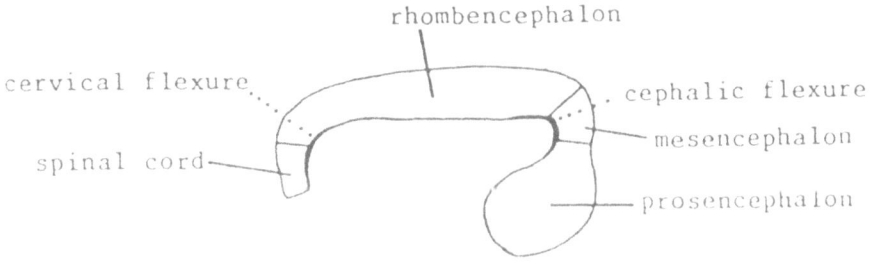

Lateral View of Developing Brain at the Fourth Week

By the sixth week of development, the <u>pontine</u> flexure occurs, causing the rhombencephalon to bend dorsally, creating the myelencephalon and the metencephalon.

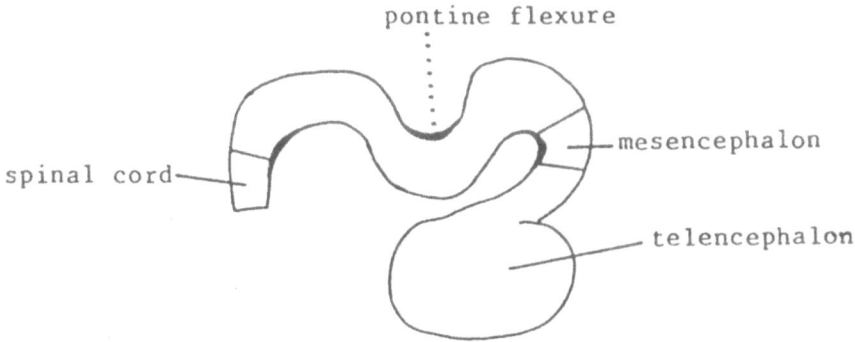

Lateral View of the Brain at Six Weeks

NOTE: Problems of normal development of the nervous system may occur, as in any other organ systems. These may frequently involve improper closure of the neural tube leading to inadequate development of the normal anatomy.

An example of this may be seen in the lumbrosacral region of the spinal cord, where myeloceles may be found (non closure of the dorsal lip of the spinal cord with intact dura), and meningomyeloceles (bulging of the meningeal coverings which contain neural elements within them). Meningomyeloceles are generally accompanied by lack of complete development of the vertebrae in this region (severe spina bifida).

The degree of severity of these types of developmental defects determines the amount of neurological deficits seen. These same maldevelopments may occur in the region of the cervical-occipital region.

Other forms of nervous system maldevelopment also occur, but space does not permit their discussion. Fortunately they are not too common.

ANATOMIC LOCALIZATION OF NEUROLOGIC LESIONS

Not uncommonly, lesions which result in changes in normal neurologic function occur in discrete locations within particular regions of the central nervous system. These lesions, particularly those resulting from interruptions in blood supply, locally occurring space occupying masses or the result of trauma, will frequently simultaneously affect different functional fiber systems and neuronal cell populations located in the damaged area. This may result in clusters of neurological signs and symptoms which many times give valuable clues as to the location of such lesions.

By contrast, other types of neurological disorder may affect only specific fiber systems (some of the demyelinating diseases for example) or specific neuronal cell populations, sparing other structures in the vicinity.

Some examples of the more recognisable lesions affecting more than one system because of their anatomic location are now presented. In the diagrams of spinal cord and brainstem cross sections, the hatched areas represent the location of lesions.

SPINAL CORD

Lesion 1.

One form of circumscribed spinal cord lesion is illustrated below, in which the affected portion of the cord is confined to one half only. Such a lesion could occur following trauma, or could also be the consequence of a compressive tumor mass. The essence of the neurological deficits is the effect on long tracts

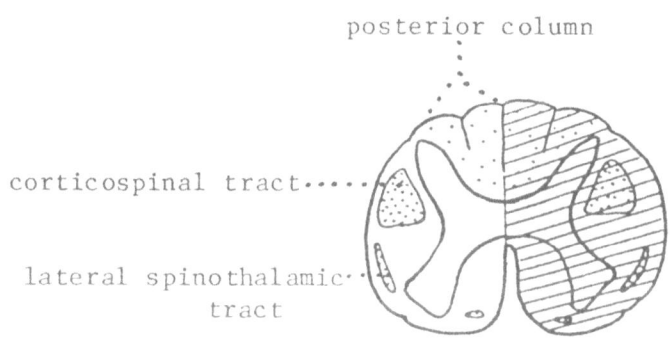

Spinal Cord with Lesion Area Shaded

travelling through the affected area, the important ones being the posterior columns and the spinothalamic tracts (both sensory) and the lateral corticospinal tract (motor). The neurological deficits resulting may be termed the Brown-Sequard Syndrome. The sensory deficits would include diminished posterior column sensation for the ipsilateral lower extremity (and also upper extremity if the cord lesion is high enough). Hence reduced joint position sense and two point discrimination etc. over the affected parts of the body. In addition, because of the involvement of the spinothalamic tracts there would also be diminished pain, thermal and tactile sensation on the <u>contralateral</u> side of the body below the lesion site. The spinothalamic system is already crossed which is why these sensations are reduced on the side contralateral to the side of the lesion. Also the sensory level resulting from this would extend up to 1 or 2 spinal segments below the site of the lesion due to the spread of incoming fibers over several segments. Motor deficits would be of the upper motor neuron type resulting from corticospinal tract involvement, and the deficits would be expressed on the <u>same</u> side of the body as the side of the lesion. Hence hyperactive reflexes, increased muscle tone and spastic paresis or paralysis, would be exhibited in the affected musculature. In addition clonus and the upgoing toe sign (Babinski response) may also be demonstrable.

NOTE: There may be a period of spinal shock occurring initially as a result of trauma, during which no reflexes may be elicited. However once this period has passed, the above described findings would be seen.

NOTE: If the unilateral lesion extends over several segments such as to compress the anterior horn cells in the are, in addition to the previously described findings there would also be evidence of <u>ipsilateral</u> lower motor neuron malfunction over the segments where the anterior horn cells were compromised. (Reduced reflexes, reduced muscle tone and flaccidity).

<u>Lesion 2.</u>

Central cord lesions in the region of the central intermediate grey matter may arise as a consequence of trauma (central cord contusion), syrinx or intrinsic cord tumor

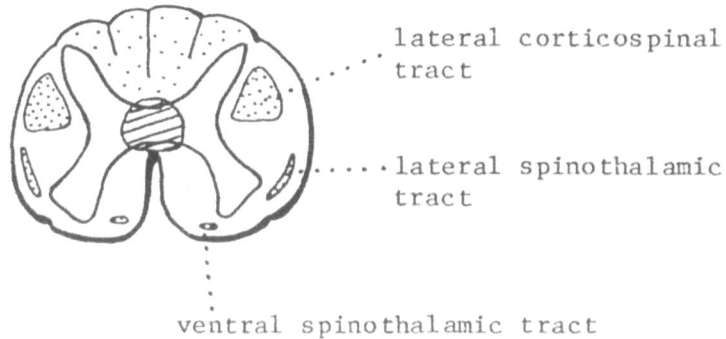

lateral corticospinal tract

lateral spinothalamic tract

ventral spinothalamic tract

Central Cord Lesion

In such circumstances the effects of these lesions may be confined to structures in the immediate area, sparing the major cord quadrants. The system which may be first compromised in these cases is the ventral white commissure, carrying the decussating fibers of the spinothalamic system. This results in a <u>bilateral</u> loss of pain and thermal sense and some diminished touch sensation over the distribution of the affected cord segments. Sensation above and below the lesion is unaffected.

NOTE: If the lesion expands in its cross sectional dimension, then
          obviously other structures may become compromised thus
          increasing the extent and complexity of the neurological
          sequellae.

BRAINSTEM

Lesion 1. Ventral caudal medulla.

Paramedian branches of the anterior spinal artery provide vascularization to this
region and, if occluded, bring about an ischaemic lesion in the area defined by the
hatched shading. The structures compromised are the medullary pyramid (above the
decussation), the exiting fibers of the ipsilateral hypoglossal nerve and, if the
lesion extends sufficiently dorsally, the ipsilateral medial lemniscus. This

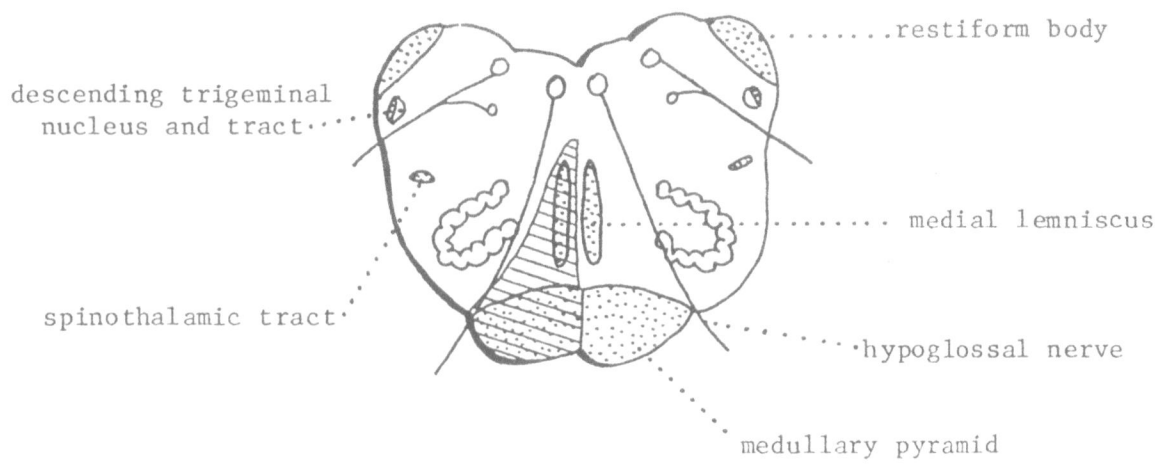

Ventral Medullary Lesion

results in motor deficits of the upper motor neuron type on the <u>contralateral</u> side
of the body below the lesion level (i.e. affects upper and lower extremities). If
the medial lemniscus is involved, in addition there will be a <u>contralateral</u> sensory
deficit involving posterior column sensation (because the lesion is <u>above</u> the
decussation of the medial lemniscus). There will also be a cranial nerve motor
deficit, as the hypoglossal nerve <u>on the side</u> of the lesion is involved. The term
<u>inferior alternating hemiplegia</u> has been used to describe this syndrome, as the
hemiplegia resulting is on one side of the body (the side opposite the lesion) and
the tongue paralysis is on the other side (same side as the lesion), causing the
tongue to deviate to the affected side on protrusion.

Lesion 2. Dorsolateral medulla

Lesions in this region, which may involve quite an array of structures, may arise as
a result of impairment of branches of the posterior inferior cerebellar artery. The
deficits resulting from this have been named <u>Wallenberg's syndrome</u>.

Deficits may include all or some of the following: 1. <u>Contralateral</u> loss of pain and thermal sensation in the body below the neck (spinothalamic tract). 2. <u>Ipsilateral</u> loss of sensation (pain and thermal sense and light touch) in the face (descending nucleus and tract of the trigeminal). 3. Difficulty in swallowing and vocalization resulting from glossopharyngeal and vagal nerve involvement. 4. <u>Ipsilateral</u> cerebellar disturbance (hemiataxia and assynergy) with restiform body compromise. 5. Eighth nerve symptoms of ipsilateral deafness (cochlear nuclei), and nausea, vomitting, vertigo and nystagmus (vestibular nuclei).

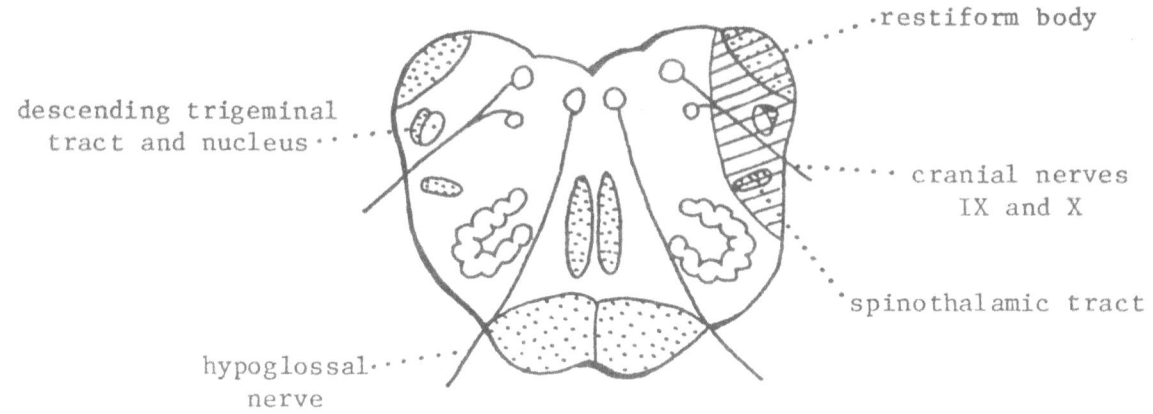

Dorsolateral Medullary Lesion

NOTE: This lesion may also include descending fibers of the autonomic sympathetic pathways in the reticular formation, in which case a <u>Horner's</u> syndrome may also be seen (ipsilateral pupillary constriction with accompanying drooping eyelid (ptosis), ipsilateral vasodilation on the face and neck with accompanying anhydrosis (absence of sweating).

NOTE: A <u>Horner's</u> syndrome may arise for entirely different reasons as a result of damage to the cervical portion of the sympathetic paravertebral chain in the neck. (<u>or</u> damage to sympathetic postganglionic fibers travelling with the internal carotid artery).

PONTINE LESIONS

Lesion 1. <u>Caudal pons</u>.

Basal pontine lesions close to the pontomedullary junction may compromise the pyramidal system and simultaneously one or more cranial nerves. One such lesion, subsequent to occlusion of branches of the anterior inferior cerebellar artery, affects the abducens nerve in addition to fibers of the pyramidal tract. This is one form of <u>middle alternating hemiplegia</u>, in which there is seen <u>ipsilateral</u> sixth nerve palsy (innability to abduct the affected eye) accompanied by <u>contralateral</u> hemiplegia in the extremities, and the findings of upper motor neuron deficit. If

the lesion extends laterally enough it may also include the fibers of the facial
nerve in addition to the previously mentioned structures. In which case <u>ipsilateral</u>
paralysis of the facial muscles of expression (above and below the eye) will be
evident as well as the involvement of the abducens nerve and the pyramidal system.
This cluster of sysmptoms has been called the <u>Millard-Gubler's</u> syndrome.

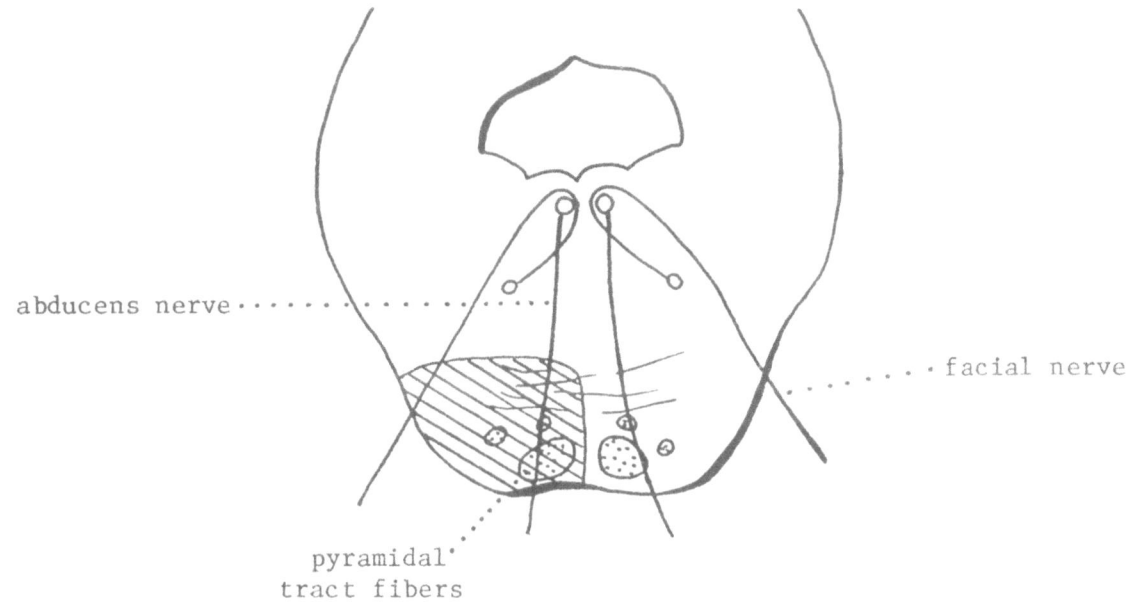

Basal Pontine Lesion

<u>Lesion 2.</u> <u>Mid Pons.</u>

Another variant of <u>middle</u> <u>alternating</u> <u>hemiplegia</u> occurs as a result of compromise of
pontine branches of the basilar artery or branches of the anterior inferior
cerebellar artery, in a more rostral location than the previously described pontine
lesion.

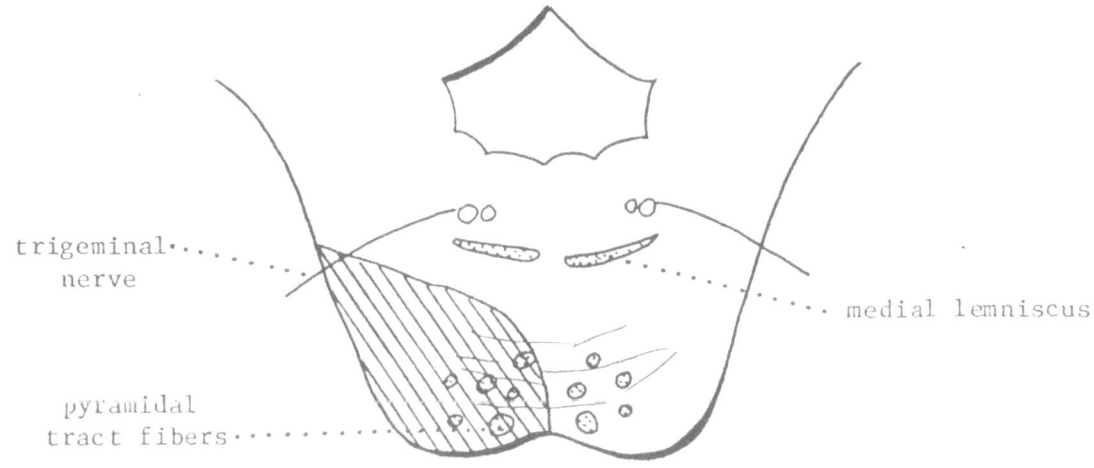

Basal Midpontine Lesion

In this instance trigeminal nerve fibers may be included in addition to the pyramidal tracts. Hemiplegia on the side <u>contralateral</u> to the lesion is now accompanied by paralysis of mastication muscles (trigeminal motor fibers) <u>ipsilaterally</u> and impairment of pain, thermal and tactile sensation ipsilaterally (trigeminal sensory fibers).

MIDBRAIN LESIONS

<u>Lesion 1.</u>

<u>Superior alternating hemiplegia</u> (Weber's syndrome) can result from unilateral lesions in the basal midbrain. In this instance the cerebral peduncle (basis pedunculi) and oculomotor nerve fibers are simultaneously involved. The resultant

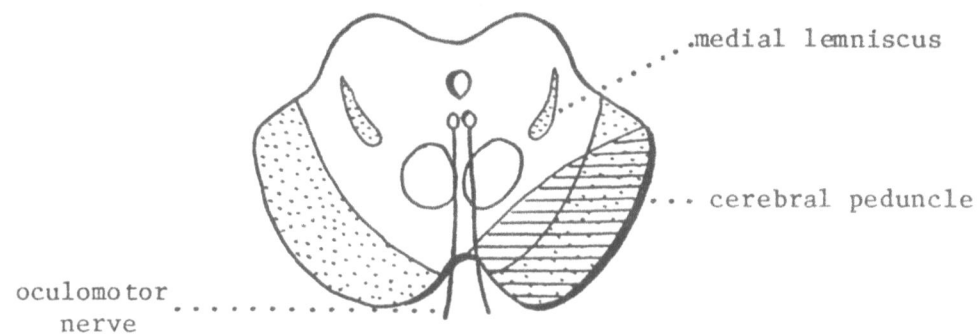

Ventrobasal Midbrain Lesion

deficits are hemiplegia on the side <u>contralateral</u> to the lesion, <u>contralateral</u> facial paralysis below the eye with <u>ipsilateral</u> third nerve paralysis. The lower facial palsy is due to the interuption of the corticobulbar fibers in the cerebal peduncle, the hemiplegia results from corticospinal involvement. Signs of oculomotor nerve palsy include dilatation of the pupil, (third nerve parasympathetics), upper eyelid ptosis and outward deviation of the eye (unopposed action of the lateral rectus muscle). Blood supply to this region is via proximal perforating branches of the posterior cerebral artery.

NOTE: All the above symptoms may be seen as a result of uncal herniation resulting from a rapidly expanding unilateral supratentoria mass (extradural hemorrhage for example) causing the ipsilateral peduncle and third nerve to be compressed against the incisural notch of the tentorium.

<u>Lesion 2.</u>

Dorsal midbrain tegmentum lesions can result in bilateral third nerve involvement plus damage to the reticular formation and ascending reticular activating system. <u>Bilateral</u> ophthalmoplegia with ptosis and dilated pupils results (third nerves). The patient is usually comatose as a result of reticular activating system malfunction.

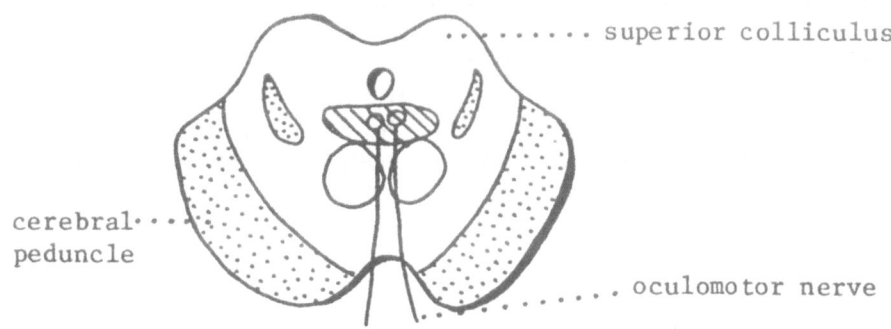

Dorsal Midbrain Tegmentum Lesion

The above described neurological problems represent just a few exaples of local
lesions affecting numerous systems and neural pathways simultaneously. Obviously
there are others which you may figure out for yourself as space limitations do not
allow their description here.

SAMPLE TEST QUESTIONS

1. A peripheral lesion of the left abducens nerve causes:

   A. Ptosis of the left upper eyelid.
   B. Abduction of the left eye in forward gaze.
   C. Absence of pupil light reflex in the left eye.
   D. Inability to gaze to the right withough moving the head.
   E. Adduction of the left eye in forward gaze.

2. The glossopharyngeal nerve is attached to the:

   A. Mesencephalon.
   B. Telencephalon.
   C. Myelencephalon.
   D. Diencephalon.
   E. Metencephalon.

3. Regarding peripheral nerve regeneration:

   A. Rate of regeneration is 10 mm per day.
   B. Regeneration only occurs in young individuals.
   C. Maximum regeneration rate is 0.2 mm per day.
   D. Maximum regeneration rate is 2 mm per day.

4. Fibers arising from the nucleus dorsalis:

   A. Carry sense of joint postions to the thalamus.
   B. Form the contralateral posterior spinocerebellar tract.
   C. Give rise to the direct arcuate fibers.
   D. Convey sense of position from the arms into the cerebellum.
   E. Form the ipsilateral posterior spinocerebellar tract.

5. All the following tracts have decussated <u>except</u> the:

   A. Medial Lemniscus.
   B. Lateral corticospinal tract.
   C. Central tegmental fasciculus.
   D. Trigeminal lemniscus.
   E. Spinal lemniscus.

6. An infarct of the left internal capsule involving the posterior limb and genu would cause:

   A. Lower right facial muscle weakness and right extremity muscle weakness.
   B. Total facial muscle paresis on the right side.
   C. Paresis of extremity and facial muscles on the left.
   D. Weakness of right extremity muscles and left lower facial muscles.
   E. Paresis of all right side facial muscles and right side extremities.

7. Structures traversing the foramen magnum include all except:

   A. Vertebral arteries.
   B. Spinal accessory nerve.
   C. Arterior spinal artery.
   D. Hypoglossal nerve.

8. Cerebellar lesions may cause all except:

   A. Nystagmus.
   B. Ataxia.
   C. Hypotonia.
   D. Dysmetria.
   E. Athetoses.

9. All the following are part of the central auditory pathway except the:

   A. Brachium of the inferior colliculus.
   B. Lateral lemniscus.
   C. Medial geniculate body.
   D. Superior olivary nucleus.
   E. Brachium of the superior colliculus.

10. Total occlusion of the anterior spinal artery would:

   A. Spare the fibers of the lateral funiculus.
   B. Affect only one half of the spinal cord.
   C. Affect lower motor neurons and spare upper motor neuron fibers.
   D. Affect sensation but spare motor mechanisms.
   E. Spare fibers in the posterior columns.

11. The glomus of the choroid plexus is located in:

   A. The roof of the third ventricle.
   B. The foramen of Magendie.
   C. The anterior limit of the temporal horn.
   D. The foramen of Lushka.
   E. The trigone region of the lateral ventricle.

12. High pitched tones are represented by activity in the:

   A. Upper part of the organ of Corti.
   B. Lateral portion of the medial geniculate body.
   C. Anterolateral portion of the transverse temporal gyri.
   D. Ventral cochlear nucleus.
   E. Lower portion of the organ of Corti.

For questions 13, 14, and 15 <u>match</u> destructive lesions in the following locations with the appropriate deficits.

    A. Lingual gyrus.
    B. Paracentral lobule.
    C. Lowermost portion of the postcentral gyrus.
    D. Angular gyrus.
    E. Inferior frontal gyrus.

13. Sensory (Wernicke's) aphasia.

14. Superior quadranopsia.

15. Weakness of musculature below the knee.

For questions 16, 17 and 18 <u>match</u> the thalamic nucleus with the appropriate afferent fiber systems:

    A. Anterior nucleus.
    B. Lateral ventral nucleus.
    C. Dorsomedial nucleus.
    D. Posterior ventral medial nucleus.
    E. Anterior ventral nucleus.

16. Trigeminal Lemniscus.

17. Mamillothalamic tract.

18. Thalamic fasciculus.

19. Compression of the oculomotor nerve could cause all <u>except</u>:

    A. Pupillary Dilatation.
    B. Ptosis of the upper eyelid.
    C. Loss of the accommodation reflex.
    D. Adduction of the affected eye.

20. Temperature sensation arising from the skin of the cheek is processed by the:

    A. Main sensory trigeminal nucleus.
    B. Posterior ventral lateral thalamic nucleus.
    C. Mesencephalic trigeminal nucleus.
    D. Descending trigeminal nucleus.

21. A lesion in the right anterior temporal lobe could cause:

    A. Left homonymous hemianopsia.
    B. Left superior quadranopsia.
    C. Left inferior quadranopsia.
    D. Right superior quadranopsia.
    E. Right homonymous hemianopsia.

22. Regions excluded from the blood brain barrier include all except:

    A. Choroid plexus
    B. Prepyriform cortex.
    C. Median eminence.
    D. Area postrema.
    E. Pineal.

For question 23, 24 and 25 match the following hypothalamic locations with the appropriate functions:

    A. Anterior hypothalamus.
    B. Posterior hypothalamus.
    C. Medial hypothalamus.
    D. Lateral hypothalamus.

23. Parasympathetic activation.

24. Production of A.D.H.

25. Lesions produce hyperphagia.

ANSWERS:

1. E.	7. D.	13. D.	19. D.	25. C.
2. C.	8. E.	14. A.	20. D.	
3. D.	9. E.	15. B.	21. B.	
4. E.	10. E.	16. D.	22. B.	
5. C.	11. E.	17. A.	23. A.	
6. A.	12. E.	18. E.	24. A.	